Metallurgy and Materials Science

a series of student texts

General Editors:

Professor R W K Honeycombe
Professor P Hancock

Light Alloys

Metallurgy of the Light Metals

I J Polmear

Professor in the Department of Materials Engineering
Monash University

Edward Arnold

© I J Polmear 1981

First published in 1981
by Edward Arnold (Publishers) Ltd
41 Bedford Square, London WCIB 3DQ

British Library Cataloguing in Publication Data

Polmear, I J
  Light alloys. – (Metallurgy and materials science).
  1. Aluminum alloys   2. Magnesium alloys
  3. Titanium alloys
  I. Title   II. Series
  669'.72   TN775

  ISBN 0–7131–2819–4

To H K Hardy

Printed in Great Britain by
Thomson Litho Ltd, East Kilbride, Scotland

General editors′ preface

Large textbooks with broad subject coverage still have their place in university teaching. However, staff and students alike are attracted to compact, cheaper books which cover well-defined parts of a subject up to and beyond final year undergraduate work. The aim of this series is to do just this in metallurgy and materials science.

The subject as taught is gradually becoming more integrated and less polarized towards metals or non-metallic materials, so we have been able to plan a group of texts concerned with basic aspects of the subject, e.g. thermodynamics, phase transformations, deformation and fracture, which will comprise one part of the series. The other books, which are concerned specifically with the main groups of materials and processing routes of engineering interest, will use the basic principles to examine and explain materials behaviour over a wide range of conditions. The aim is not to cover each subject in the greatest depth, but to provide the student with a compact treatment as a springboard to further detailed studies, when he has chosen his particular field of work after graduation. Adequate general references will be provided for further study. The books are aimed not only at metallurgists and materials scientists, but also towards engineers and scientists wishing to know more about the structure and properties of engineering materials.

1980   RWKH
                                                                          PH

Preface

The fact that the light metals aluminium, magnesium and titanium have traditionally been associated with the aerospace industries has tended to obscure their growing importance as general engineering materials. For example, aluminium is now the second most widely-used metal and production during the next two decades is predicted to expand at a rate greater than that for all other structural metals. Titanium, which has a unique combination of properties that have made its alloys vital for gas turbine engines, is now finding many applications in aircraft structures and in the chemical industry.

Light alloys have never been the subject of a single book. Moreover, although the general metallurgy of each class of light alloys has been covered in individual texts, the most recent published in English appeared some time ago – aluminium alloys in 1970, magnesium alloys in 1966 and titanium alloys in 1956. Many new developments have occurred in the intervening periods and important new applications are planned, particularly in transportation. Thus it is hoped that the appearance of this first text is timely.

In preparing the book I have sought to cover the essential features of the metallurgy of the light alloys. Extraction of each metal is considered briefly in Chapter one, after which the casting characteristics, alloying behaviour, heat treatment, properties, fabrication and major applications are discussed in more detail. I have briefly reviewed the physical metallurgy of aluminium alloys in Chapter two although the general principles also apply to the other metals. Particular attention has been devoted to microstructure/property relationships and the role of individual alloying elements, which provides the central theme. Special features of light alloys and their place in general engineering are highlighted although it will be appreciated that it has not been possible to pursue more than a few topics in depth.

The book has been written primarily for students of metallurgy and engineering although I believe it will also serve as a useful guide to both producers and users of light alloys. For this reason, books and articles for further reading are listed at the end of each chapter and are augmented by the references included with many of the figures and tables.

The book was commenced when I was on sabbatical leave at the Joint Department of Metallurgy at the University of Manchester Institute of Science and Technology and University of Manchester, so that thanks are due to Professor K. M. Entwistle and Professor E. Smith for the generous facilities placed at my disposal. I am also indebted for assistance given by the Aluminium Development Council of Australia and to many associates who have provided me with advice and information. In this regard, I wish particularly to mention the late Dr E. Emley, formerly of The British Aluminium Company Ltd; Dr C. Hammond, The University of Leeds; Dr M. Jacobs, TI Research Laboratories; Dr D. Driver, Rolls-Royce Ltd; Dr J. King and Mr W. Unsworth, Magnesium Elektron Ltd; Mr R. Duncan, IMI Titanium; Dr D. Stratford, University of Birmingham; Dr C. Bennett, Comalco Australia Ltd; and my colleague Dr B. Parker, Monash University. Acknowledgement is also made to publishers, societies and individuals who have provided figures and diagrams which they have permitted to be reproduced in their original or modified form.

Finally I must express my special gratitude to my secretary Miss P. O'Leary and to Mrs. J. Colclough of the University of Manchester who typed the manuscript and many drafts, as well as to Julie Fraser and Robert Alexander of Monash University who carefully produced most of the photographs and diagrams.

Melbourne   IJP
1980

Contents

x   *Contents*

1
The light metals

1.1 General introduction

The term 'light metals' has traditionally been given to both aluminium and magnesium because they are frequently used to reduce the weight of components and structures. On this basis, titanium also qualifies and beryllium should be included although it is little used and will not be considered in detail in this book. These four metals have relative densities ranging from 1.7 (magnesium) to 4.5 (titanium) which compare with 7.9 and 8.9 for the older structural metals, iron and copper, and 22.6 for osmium, the heaviest of all metals. Nine other elements that are classified as metals are lighter than titanium but, with the exception of boron in the form of strong fibres contained in a suitable matrix, none is used as a base material for structural purposes. The alkali metals lithium, potassium, sodium, rubidium and caesium, and the alkaline earth metals calcium and strontium are too reactive, whereas scandium is comparatively rare.

The property of lightness has led to the association of the light metals with transportation and more especially with aerospace which has provided great stimulus to the development of alloys during the last 30 years. Strength:weight ratios have thus been a dominant consideration and these are particularly important in engineering design when parameters such as stiffness or resistance to buckling are involved. For example, the stiffness of a simple rectangular beam is directly proportional to the product of the elastic modulus and the cube of the thickness. The significance of this relationship is illustrated by the nomograph shown in Fig. 1.1 which allows the weights of similar beams of different metals and alloys to be estimated for equal values of stiffness. An iron, or steel, beam weighing 10 kg will have the same stiffness as beams of equal width and length weighing 7 kg in titanium, 4.9 kg in aluminium, 3.8 kg in magnesium, and only 2.2 kg in beryllium. The Mg-Li alloy is included because it is the lightest (relative density 1.35) structural alloy that is available commercially. Comparative stiffnesses for equal weights of a similar beam increase in the ratios 1:2.9:8.2:18.9 for steel, titanium, aluminium, and magnesium respectively.

Concern with aspects of weight saving should not obscure the fact that light metals possess other properties of considerable technological importance, e.g. the high corrosion resistance and electrical and thermal conductivities of aluminium, the machinability of magnesium, and extreme

1

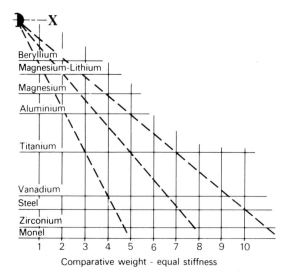

Comparative weight - equal stiffness

Fig. 1.1 Nomograph allowing the comparative weights of different metals or alloys to be compared for equal levels of stiffness. These values can be obtained from the intercepts which a line drawn from point X makes with lines representing the different metals or alloys (courtesy Brooks and Perkins Inc.)

corrosion resistance of titanium. Comparisons of some physical properties are made in Table 1.1.

The estimated crustal abundance of the major chemical elements is given in Table 1.2 which shows that the light metals aluminium, magnesium, and titanium are first, third, and fourth in order of occurrence of the structural metals. It can also be seen that the traditional metals copper, lead, and zinc are each present in amounts less than 0.10 %. Estimates are also available for the occurrence of metals in the ocean which is the major commercial source of magnesium. Sea water contains 0.13 % of this metal so that 1.3 million tonnes are present in each km^3, which is approximately equivalent to one-quarter of the total world consumption of magnesium up until the present time. Overall, the reserves of the light metals are adequate to cope with anticipated demands for some centuries to come. The extent to which they will be used would seem to be controlled mainly by their future costs relative to competing materials such as steel and plastics, as well as the availability of electrical energy that is needed for their extraction from minerals.

Trends in the consumption of various metals and plastics are shown in Fig. 1.2 and it is clear that the light metals are very much materials of the 20th Century. Between 1900 and 1950, the annual world production of aluminium increased 250 times from around 6000 tonnes to 1.5 million tonnes. A further eight-fold increase took place during the next quarter century when aluminium surpassed copper as the second most used metal.

Table 1.1　Some physical properties of pure metals (from Smithells, C. J., *Metals Reference Book*, 5th edition, Butterworths, London, 1976)

Property	Unit	Al	Mg	Ti	Be	Fe	Cu
Melting point	°C	660	650	1678	1287	1535	1083
Relative density (*d*)		2.70	1.74	4.51	1.85	7.87	8.96
Elastic modulus (*E*)	GPa	70	45	120	295	211	130
Specific modulus (*E/d*)		26	26	26	160	27	14
Mean specific heat 0–100°C	$J\,kg^{-1}\,K^{-1}$	917	1038	528	2052	456	386
Thermal conductivity 20–100°C	$W\,m^{-1}\,K^{-1}$	238	156	26	194	78	397
Coefficient of thermal expansion 0–100°C	$10^{-6}\,K^{-1}$	23.5	26.0	8.9	12.0	12.1	17.0
Electrical resistivity at 20°C	μ ohm cm^{-1}	2.67	4.2	54	3.3	10.1	1.69

Note:　Conversion factors for SI and Imperial units are given in the Appendix

Table 1.2   Crustal abundance of major chemical elements (from Stanner, R. J. L., *American Scientist*, **64**, 258, 1976)

Element	% by weight
Oxygen	45.2
Silicon	27.2
Aluminium	8.0
Iron	5.8
Calcium	5.06
Magnesium	2.77
Sodium	2.32
Potassium	1.68
Titanium	0.86
Hydrogen	0.14
Manganese	0.10
Phosphorus	0.10
Total	99.23

Production of primary aluminium in 1979 was 15.1 million tonnes as compared with 9.4 million tonnes of copper. Consumption of aluminium becomes significantly greater if calculated on a volumetric basis, but it should be noted that iron and steel amount to more than 90 wt % of all metal produced. Production of magnesium has remained close to 250 000 tonnes per annum in recent years and equals only 2 % that of aluminium. Consumption of titanium has increased at an average annual rate of 8 to 10 % in recent years and the annual production for 1979 was 80 000 tonnes.

Prices of metals may vary depending upon factors such as supply and demand. For example, the price of copper ingot per tonne in the United States ranged from 2.1 to 1.2 times that for aluminium ingot during the years 1974 to 1978. Similar ratios for magnesium with respect to aluminium were 1.2 to 1.9 for the same period. Direct comparisons between aluminium and steel are more difficult to make although it may be noted, for example, that aluminium alloy sheet is normally three to four times more expensive than mild steel sheet of the same weight. This differential becomes much less with products for which volume or area are prime considerations.

The leading world producers of aluminium are the United States (4.7 million tonnes per annum in 1977), Soviet Union (2 million tonnes) and Japan (1.5 million tonnes), with the Soviet Union having the largest smelter in the world (rated output 600 000 tonnes per annum). Current predictions are that the production of aluminium will expand at an average annual rate of 4.5 % during the next two decades compared with 5 % for plastics, 2.4 % for steel and 2.3 % for copper. Of the major structural materials, only the consumption of aluminium and plastics is expected to expand at a rate greater than the predicted rate of growth in the Gross National Products of the developed countries. At the same time, large increases in energy costs

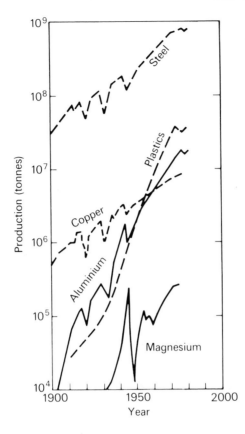

Fig. 1.2 World production figures for various metals and plastics (from Crowther, J., *Metals and Materials*, **17**, 270, 1973)

are already causing the closure of some aluminium smelters in the United States and Japan. Production of aluminium is being transferred to countries where cheaper energy is available, one example being Australia where the output is expected to treble to around 750 000 tonnes by 1982. In effect, countries such as Australia will be exporting energy in the form of aluminium which has led to the notion of this metal being an 'energy bank'.

Aluminium is used in four major areas in most countries. These are building and construction, containers and packaging, transportation, and electrical conductors which are commonly ranked in this order. The major growth market in the future is expected to be transportation as more aluminium is used in motor cars. In this regard, it is predicted that this amount will double in American cars during the five-year period 1979 to 1984.

More attention is being given to the recycling of materials and the incentive for this is particularly strong in the case of aluminium because remelting of scrap requires only 5 % of the energy needed to produce the same weight of primary aluminium from the ore bauxite. Currently, the ratio of secondary (scrap) to primary aluminium that is used is estimated to be around 25 % which is well below what is recovered with steel and copper. Aluminium alloys do present a special problem, however, because they cannot be refined. Remelting thus tends to downgrade the alloys so that they are used almost exclusively for foundry castings which, in turn, are limited in the amount they can absorb. Thus, there is a need to accommodate more aluminium scrap in cast billets to be used for producing wrought alloys. Since more than 60 % of scrap originates from fabricating processes, some recycling into wrought products has been possible by retaining the scrap within a closed circuit. Packaging provides another major source of scrap which is now largely discarded and there is a trend, particularly in the United States, to collect and recycle the all-aluminium beverage cans to produce the sheet used for can stock.

More than half the magnesium produced is used as alloying additions to aluminium and nodular (SG) cast iron. The remainder is used mainly for castings in the aerospace and general transport industries and in this respect magnesium alloys with an interesting variety of other metals including zinc, zirconium, thorium, rare earth metals, silver and lithium. Less use is made of wrought products although the selection of extruded magnesium alloy fuel element cans for the British Magnox gas cooled nuclear reactors is one notable example.

Titanium was not produced in quantity until the late 1940s when its relatively low density and high melting point (1678° C) made it uniquely attractive as a potential replacement for aluminium for the skin and structure of high-speed aircraft subjected to aerodynamic heating. Liberal military funding was provided in the decade 1947–57 and one of the major metallurgical investigations of all time was made of titanium and its alloys. It is estimated that $400 million was spent in the United States during this period and one firm examined more than 3000 alloys. One disappointing result was that titanium alloys showed relatively poor creep properties bearing in mind their very high melting points. This factor, together with a sudden change in emphasis from manned aircraft to guided weapons, led to a slump in interest in titanium in 1957–58. Since then, selection of titanium alloys for engineering uses has been made on the more rational bases of cost-effectiveness and the uniqueness of certain properties. The high specific strength of titanium alloys when compared with other light alloys, steels, and nickel alloys is apparent in Fig. 1.3. The fact that this advantage is maintained to around 500° C has led to the universal acceptance of certain titanium alloys for critical gas turbine components, and applications in the aerospace industry account for some 80 % of titanium that is produced. Most of the remainder is used in the chemical industry.

Beryllium has some remarkable properties. Its specific modulus is nearly an order of magnitude greater than for aluminium, steel and other common

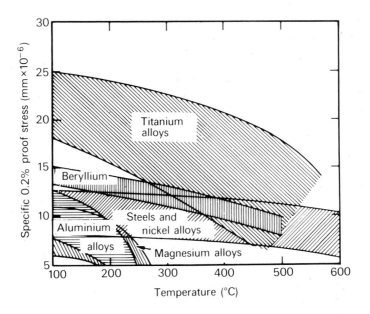

Fig. 1.3 Relationship of specific 0.2% proof stress (ratio of proof stress to relative density) with temperature for light alloys, steels and nickel alloys

materials, it has a relatively high melting point, and its capture cross-section for neutrons is lower than any other metal. These properties have stimulated considerable interest by aerospace and nuclear industries. A design study using beryllium as the major structural material for a supersonic transport aircraft has indicated possible weight savings of up to 50% for the components for which it could be used. Interest in nuclear engineering has arisen because of its potential use as a fuel element can in power reactors. Despite much research in several countries, however, negligible use is made of beryllium mainly due to cost, its inherently low ductility at ambient temperatures and the fact that the powdered oxide is extremely toxic to some people.

1.2 Production of aluminium

Although aluminium was isolated in small quantities early in the 19th Century, it remained an expensive curiosity until 1886 when independent discoveries by Hall in America and Héroult in France led to the development of an economic method for its electrolytic extraction. Since then, the emergence of aluminium as a practical, commercial metal has relied primarily on the availability of bulk quantities of electricity at economic prices.

Aluminium is obtained from bauxite which is the name given to ores usually containing 40 to 60 % hydrated alumina together with impurities such as iron oxides, silica and titania. The name originates from Les Baux, the district in Provence, France where the ore was first mined. Bauxite is formed by the surface weathering of aluminium-bearing rocks such as granite and basalt under tropical conditions and the largest known reserves are found in Northern Australia, Guyana and Brazil. Australia (30 %), Jamaica (15 %) and Guyana (12 %) are the current major suppliers. High grade bauxites with low silica contents are expected to be depleted in 20–30 years time. When it becomes necessary to use high-silica bauxites the ore will need to be processed to remove this impurity, probably by froth flotation, and this will introduce an additional cost penalty.

Immense amounts of aluminium are also present in clays, shales and other minerals but it is difficult and uneconomic to extract the metal from these resources. An exception is in the Soviet Union where high-grade bauxite is unavailable and some plants are located in areas remote from sources of this ore.

Production of aluminium from bauxite involves two distinct processes which are often operated at quite different locations. First, pure alumina (Al_2O_3) is extracted from bauxite almost exclusively by the Bayer process which, essentially, involves digesting crushed bauxite with strong caustic soda solution at temperatures up to 240° C. Most of the alumina is extracted leaving an insoluble residue known as red mud, consisting largely of iron oxide and silica which is removed by filtration. After cooling, the liquor is seeded with crystals of alumina trihydrate to reverse the chemical reaction, the trihydrate being precipitated and the caustic soda recycled. The whole process can be represented by the chemical reaction:

$$Al(OH)_3 + NaOH \rightleftarrows NaAlO_2 + 2H_2O$$

The alumina trihydrate is then calcined in a rotary kiln at 1200° C to remove water of crystallization and alumina is produced as a fine powder.

Alumina has a high melting point (2040° C) and is a poor conductor of electricity. The key to the successful production of aluminium lies in dissolving the alumina in molten cryolite (Na_2AlF_6) and a typical electrolyte contains 80 to 90 % of this compound and 2 to 8 % alumina together with additives such as aluminium and calcium fluorides. Cryolite was first obtained from relatively inaccessible sources such as Greenland but is now made synthetically.

Fig. 1.4 depicts a section from a multi-anode production cell. The outer casing consists of a brick-lined, rectangular steel box which contains baked carbon or graphite blocks that serve both as the cathode and to collect the molten aluminium. The anodes are made from pre-baked carbon blocks that dip into the electrolyte and are gradually consumed in the reaction. Alternatively, a single large (Soderberg) electrode is used in which the anode carbon paste is added *in situ* and baked by the waste heat from the cell. The bath runs at around 1000° C and cells are arranged in series in so-called potlines. Current loadings are now as high as 100 000 to 200 000 A

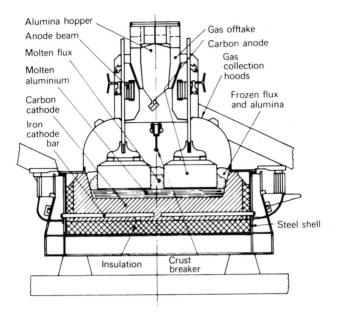

Alumina hopper
Anode beam
Molten flux
Molten aluminium
Carbon cathode
Iron cathode bar

Gas offtake
Carbon anode
Gas collection hoods
Frozen flux and alumina
Steel shell
Insulation
Crust breaker

Fig. 1.4 Multi-anode electrolytic cell for extracting aluminium from alumina (courtesy Australian Aluminium Development Council)

with a voltage drop of 5 V across each cell. The aluminium is extracted by tapping or syphoning at intervals and alumina is added as required.

The exact mechanism for the electrolytic reaction in the cell is uncertain but it is probable that the current carrying ions are Na^+ and one or more complex ions such as $AlOF_3^{--}$. At the cathode, sodium is liberated and reacts with aluminium fluoride, or one of the complex fluorides, to produce aluminium while, at the anode, the complex ions dissociate to liberate oxygen which forms CO_2. The overall reaction can be written as follows:

$$2Al_2O_3 + 3C \rightarrow 4Al + 3CO_2$$

Fig. 1.5 shows a flow diagram for the raw materials needed to produce one tonne of aluminium. It will be seen that 3.5 to 4 tonnes of bauxite are needed to produce 1.9 tonnes of alumina which in turn yields one tonne of aluminium. Significant quantities of other materials are also consumed including approximately 0.3 tonnes of fuel oil and a net 0.45 tonnes of carbon. As mentioned earlier, the most critical factor is the consumption of electricity which, despite continual refinements to different parts of the process, still amounts to a minimum of 13 000 to 14 000 kWh of electricity for each tonne of aluminium extracted from alumina. Overall, the total energy requirement to produce one tonne of primary aluminium from bauxite in the ground is estimated to be between 70 000 and 75 000 kWh (thermal), where the total energy required has been converted back to an

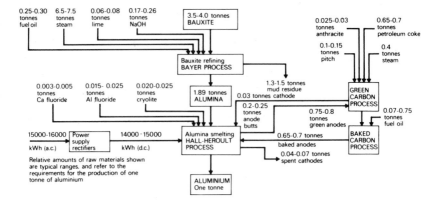

Fig. 1.5 Flow diagram for integrated production of aluminium from bauxite (courtesy Australian Aluminium Development Council)

equivalent amount of fossil fuel by assuming 1 kWh (electricity) = 3 kWh (thermal). This reduces to about 30 000 kWh (thermal) per unit volume of aluminium, but is still much greater than the estimated 13 000 to 16 000 kWh (thermal) of energy required to produce 1 tonne of steel from iron ore in the ground to the finished product.

Alternative methods for extracting aluminium have been investigated but only one new process has been developed to a commercial scale. This is the Alcoa smelting process which commenced operation in Texas in 1976 with an initial capacity of 13 500 tonnes of primary aluminium per year. This new process also uses alumina as a starting material which is combined with chlorine in a reactor to produce aluminium chloride. The chloride is treated electrolytically in an enclosed cell to produce aluminium and chlorine, the latter being recycled back to the reactor. The process is continuous and, since it is completely enclosed, working conditions are improved. Moreover, environmental problems with fluoride emissions are avoided because cryolite is not used in the electrolytic cell. The process is still being fully evaluated but reports so far suggest that a 30 % saving in electricity is possible. Another advantage is that the plant occupies less than half the land area per tonne capacity that is needed with the conventional process. A major problem at present is the provision of materials of construction that can resist attack by chlorine over long periods of time.

1.3 Production of magnesium

Magnesium was first extracted by Humphrey Davy in 1808 but it remained little more than a curiosity until the early part of the present century when the advent of heavier-than-air flying machines stimulated interest in all lightweight constructional materials. Magnesium compounds are found in abundance, both as solid deposits and in solution in surface waters of the earth.

The most common minerals found in the earth's crust are the carbonates dolomite ($MgCO_3 CaCO_3$) and magnesite ($MgCO_3$). The oxide mineral brucite ($MgO.H_2O$) is somewhat rarer, as are the chlorides of which carnallite ($MgCl_2 KCl.6H_2O$) is one example. Concentrated aqueous solutions in the form of brine deposits occur at several places in the world but by far the largest deposit is in the world's oceans. The magnesium content of the ocean is 0.13 % which is a smaller concentration than the lowest grade deposit of minerals mentioned above. However, the supply is virtually unlimited and the uniformity of the magnesium content permits highly standardized procedures to be used for its extraction. Sea water is now the main source of magnesium.

Although many processes have been developed to produce metallic magnesium, only three methods are now used. The earliest method, and the one that accounts for 80–90 % of present output, involves the electrolytic reduction of magnesium chloride. In the other two processes, dolomite is directly reduced by ferrosilicon at high temperatures. One is known as the Pidgeon process in which the reaction is carried out in the solid state. It is only economic under comparatively rare conditions where there is a natural site advantage. The other is the more recent Magnethérme process developed in France which is carried out at much higher temperatures so that the reaction mixture is liquid. It is assuming increasing importance.

Two electrolytic processes are currently in use to produce magnesium which differ mainly in the degree of hydration of the $MgCl_2$ and the cell characteristics. One was pioneered by the German firm IG Farbenindustrie early in this century and is now used by the Norwegian firm Norsk Hydro which is the main European producer of this metal. Known originally as the IG Process, it uses dry MgO derived from minerals or sea water which is briquetted with a reducing agent, e.g. powdered coal, and with $MgCl_2$ solution. The briquettes are lightly calcined and then chlorinated at around 1100°C to produce molten anhydrous $MgCl_2$ which is fed directly to the electrolytic cells operating at about 750°C. Other chlorides such as NaCl and $CaCl_2$ are added to improve the conductivity, viscosity and density of the electrolyte. Each cell has graphite anodes and cast steel cathodes suspended opposite one another. The magnesium collects as droplets on the faces of the cathodes and rises to the surface of the electrolyte, whereas chlorine liberated at the anodes is recycled to produce the initial $MgCl_2$ cell feed.

The second electrolytic process was developed by the Dow Chemical Company and is used at the world's largest magnesium plant at Freeport, Texas, which extracts $MgCl_2$ from sea water. Magnesium is precipitated as the hydroxide by the addition of lime and then dissolved in HCl. The solution is subsequently concentrated and dried although the process stops short of complete dehydration of the $MgCl_2$ which is then available as a cell feed. In contrast to the IG-Norsk Hydro process, the cells require external heat with the steel-containing pot serving as the cathode. Cell currents are around 60 000 A which is more than double that normally used in the other process. The energy consumed per kg of magnesium is around 17.5 kWh using either process.

Production of magnesium by direct thermal reduction of calcined dolomite with ferrosilicon proceeds according to the simplified equation:

$$2CaO\,MgO + Si \rightarrow 2Mg + (CaO)_2\,SiO_2$$

Briquettes of the reactants are heated to about 1150°C in steel retorts under vacuum and the magnesium evolved is deposited in a simple tubular condenser inserted in the cold end of the retort. One advantage of the process is the less rigid requirement placed on the purity of the raw materials although, as mentioned earlier, it is generally uneconomic when compared with the electrolytic methods.

1.4 Production of titanium

The existence of titanium was first recognized in 1791 by William McGregor, an English clergyman and mineralogist, who detected the oxide of a new metal in the mineral ilmenite ($FeO\,TiO_2$). An impure sample of titanium was first isolated in 1825 but it was not produced in any quantity until 1937 when Kroll, in Luxembourg, reacted $TiCl_4$ with molten magnesium under an atmosphere of argon. This opened the way to the industrial exploitation of titanium and the essential features of the process are as follows:

(i) Briquette TiO_2 with coke and tar and chlorinate at 800°C to promote the reaction: $TiO_2 + 2Cl_2 + 2C \rightarrow TiCl_4 + 2CO$.

(ii) Purify $TiCl_4$ by fractional distillation.

(iii) Reduce $TiCl_4$ by molten magnesium or sodium under an argon atmosphere, one reaction being: $TiCl_4 + 2Mg \rightarrow Ti + 2MgCl_2$.

Titanium forms as an impure sponge around the walls of the reduction vessel and is removed periodically. The sponge produced by reacting with magnesium must be purified by leaching with dilute HCl and/or distilling off the surplus $MgCl_2$ and magnesium. The use of sodium has the advantages that leaching is more efficient and the titanium sponge is granular, making it easier to compact for the subsequent melting process.

The mineral rutile (TiO_2) is the most convenient source of titanium and is found mainly in beach sands along the eastern coast of Australia and in estuaries in Sierra Leone. Most titanium metal is extracted from rutile although the much more plentiful, but more complex mineral ilmenite will become the major source in the future.

The controlling factor in refining titanium sponge is the metal's high reactivity with other elements, notably its affinity for oxygen, nitrogen, hydrogen, and carbon. As shown in Table 1.3, the solubility of these interstitial elements in titanium is greater by several orders of magnitude than in other commonly used metals. Since quite small amounts of these elements adversely affect the ductility and toughness of titanium, it is clearly impossible to melt in air or in a normal crucible because the metal

Table 1.3 Solubility at room temperature of O, N, C and H in titanium, iron and aluminium (from Morton, P. H., *The Contribution of Physical Metallurgy to Engineering Practice*, Rosenhain Centenary Conference, The Royal Society, 1976)

| Metal | Interstitial element | | | |
	Oxygen	Nitrogen	Carbon	Hydrogen
Titanium	14.5 wt%	~20 wt%	0.5 wt%	~100 ppm
Iron	~1 ppm	<5 ppm	100 ppm	<1 ppm
Aluminium	<1 ppm	<1 ppm	<1 ppm	<1 ppm

ppm = parts per million

will adsorb gases and react with any known oxide or carbide refractory. Accordingly, a radically new method of melting had to be devised leading to what is known as the consumable-electrode arc furnace (Fig. 1.6).

Melting is carried out in a copper crucible cooled internally by circulating water or a liquid sodium-potassium eutectic. Heat is generated by a direct current arc that is struck between an electrode of titanium to be

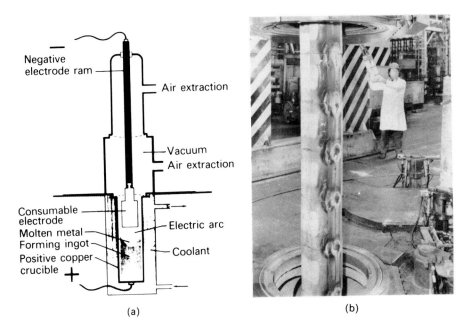

(a)   (b)

Fig. 1.6a Consumable-electrode arc furnace for melting and refining titanium; b, consumable electrode made by welding together blocks of compressed titanium sponge (courtesy T. W. Farthing)

melted and a starting slug of this material contained in the crucible. An advantage of the liquid alloy coolant is that it does not react with titanium should the electric arc perforate the crucible, whereas water and steam can cause an explosion. The electrode is usually made from welded blocks of compressed titanium sponge to which alloying elements are incorporated in powder form (Fig. 1.6). The entire arrangement is encased in a vessel which can be evacuated and to which an inert gas, e.g. argon, can be introduced. The capacity of the furnace is increased by using a retractable hearth and ingots weighing 2 to 3 tonnes can be produced. Double melting is normally carried out to improve the homogeneity of ingots.

Overall the energy consumed in making pure titanium sponge is greater than is required for producing any other common metal in ingot form. For example, it is currently about 70 % higher than that needed for extracting an equal weight of aluminium.

Further reading

Varley, P. C., *The Technology of Aluminium Alloys*, Newnes-Butterworths, London, 1970

Emley, E. F., *Principles of Magnesium Technology*, Pergamon, London, 1966

World Bureau of Metal Statistics, *World Metal Statistics*, London, 1979

Crowther, J., Materials competition – the statistical background, *Metals and Materials*, **7**, 270, 1973

Netschert, B. C., The future availability of raw materials, *J. Metals*, **30**, No. 8, 12, 1978

Burte, H. M., The nature of competition between emerging materials, *J. Metals*, **30**, No. 8, 17, 1978

Dowding, M., The world of metals, *Metals and Materials*, July, p. 27, 1978

Dowsing, R. J., Aluminium second only to steel in worldwide engineering use, *Metals and Materials*, January, p. 20, 1977

Moore, J. J., Recycling of non-ferrous metals, *Int. Met. Rev.*, **23**, 241, 1978

Farthing, T. W., Introducing a new material – the story of titanium, *Proc. Inst. Mech. Eng.*, **191**, 159, 1977

Dowsing, R. J., Spotlight on titanium, *Metals and Materials*, July/August, p. 27; September, p. 31; December, p. 43, 1979

Altenpohl, D., *Materials in World Perspective*, Springer-Verlag, Berlin, 1980

2
Physical metallurgy of aluminium alloys

Although most metals will alloy with aluminium, comparatively few have sufficient solid solubility to serve as major alloying additions. Of the commonly used elements, only zinc, magnesium (both greater than 10 atomic %)[†], copper and silicon have significant solubilities (Table 2.1). However, several other elements with solubilities below 1 atomic % confer important improvements to alloy properties. Examples are some of the transition metals, e.g. chromium, manganese and zirconium, which are

Table 2.1 Solid solubility of elements in aluminium (from Van Horn, K. R. (ed), *Aluminum*, Volume 1, American Society of Metals, Ohio, 1967; Mondolfo, L. F., *Aluminium Alloys: Structure and Properties*, Butterworths, London, 1976)

Element	Temperature (°C)	Maximum solid solubility	
		(wt %)	(at %)
Cadmium	649	0.4	0.09
Cobalt	657	<0.02	<0.01
Copper	548	5.65	2.40
Chromium	661	0.77	0.40
Germanium	424	7.2	2.7
Iron	655	0.05	0.025
Lithium	600	4.2	16.3
Magnesium	450	17.4	18.5
Manganese	658	1.82	0.90
Nickel	640	0.04	0.02
Silicon	577	1.65	1.59
Silver	566	55.6	13.8
Tin	228	∼0.06	∼0.01
Titanium	665	∼1.3	∼0.74
Vanadium	661	∼0.4	∼0.21
Zinc	443	70	28.8
Zirconium	660.5	0.28	0.08

Note:
(i) Maximum solid solubility occurs at eutectic temperatures for all elements except chromium, titanium, vanadium, zinc and zirconium for which it occurs at peritectic temperatures.
(ii) Solid solubility at 20°C is estimated to be approximately 2 wt % for magnesium and zinc, 0.1 to 0.2 wt % for germanium, lithium and silver and below 0.1 % for all other elements.

[†] Unless stated otherwise, alloy compositions are quoted in weight percentages.

15

used primarily to form compounds that control grain structure. With the exception of hydrogen, elemental gases have no detectable solubility in either liquid or solid aluminium. Apart from tin which is sparingly soluble, maximum solid solubility in binary aluminium alloys occurs at eutectic and peritectic temperatures. Sections of typical eutectic and peritectic binary phase diagrams are shown in Figs. 2.1 and 2.2.

Aluminium is a soft metal and the fact that high strength:weight ratios can be achieved in certain alloys arises because they show a marked response to age or precipitation hardening. It is, therefore, desirable to present a brief review of the essential principles associated with this phenomenon before considering specific alloy systems. These remarks also apply generally to those magnesium and titanium alloys in which precipitation occurs.

2.1 Principles of age hardening

2.1.1 Decomposition of supersaturated solid solutions

The basic requirement for an alloy to be amenable to age hardening is a decrease in solid solubility of one or more of the alloying elements with decreasing temperature. Heat treatment normally involves:
(i) Solution treatment at a relatively high temperature within the single phase region, e.g. A in Fig. 2.1, to dissolve the alloying elements.
(ii) Rapid cooling or quenching, usually to room temperature, to ob-

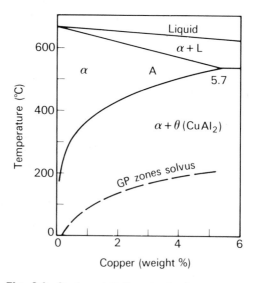

Fig. 2.1   Section of Al-Cu eutectic phase diagram. The position of GP zones solvus is also shown

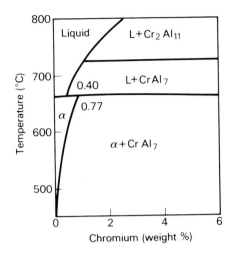

Fig. 2.2   Section of Al-Cr peritectic phase diagram

tain a supersaturated solid solution (SSSS) of these elements in aluminium.

(iii) Controlled decomposition of the SSSS to form a finely dispersed precipitate, usually by ageing for convenient times at one and sometimes two intermediate temperatures.

The complete decomposition of a SSSS is usually a complex process which may involve several stages. Typically, Guinier-Preston (GP) zones and an intermediate precipitate may be formed in addition to the equilibrium phase. The GP zones are ordered, solute-rich clusters of atoms which may be only one or two atom planes in thickness. They retain the structure of the matrix and are coherent with it, although they usually produce appreciable elastic strains (Fig. 2.3). Their formation requires movements of atoms over relatively short distances so that they are very finely dispersed in the matrix with densities which may be as high as 10^{17} to 10^{18} cm^{-3}. Depending upon the alloy system, the rate of nucleation and the actual structure may be greatly influenced by the presence of excess vacant lattice sites that are also retained by quenching.

The intermediate precipitate is normally much larger in size than a GP zone and is only partly coherent with the lattice planes of the matrix. It has a definite composition and crystal structure which may differ only slightly from those of the equilibrium precipitate. In some alloys, the intermediate precipitate may be nucleated from, or at, the sites of stable GP zones. In others this phase nucleates heterogeneously at lattice defects such as dislocations (Fig. 2.4). Formation of the final equilibrium precipitate involves complete loss of coherency with the parent lattice. It only forms at relatively high ageing temperatures and, because it is coarsely dispersed, little hardening results.

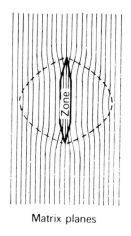

Matrix planes

Fig. 2.3 Representation of the distortion of matrix lattice planes near to the coherent GP zone (from Nicholson, R. B. *et al.*, *J. Inst. Metals*, **87**, 429, 1958–59)

Maximum age hardening in an alloy occurs when there is present a critical dispersion of GP zones or an intermediate precipitate, or both.

2.1.2 The GP zones solvus

An important concept is that of the GP zones solvus which may be shown as a metastable line in the equilibrium diagram (Fig. 2.1). It defines the

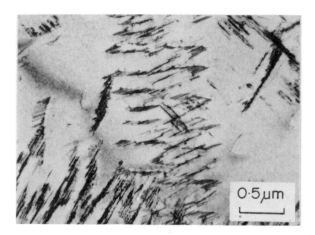

0·5μm

Fig. 2.4 Transmission electron micrograph showing the rods of the S-phase (Al$_2$CuMg) precipitated heterogeneously on dislocation lines. The alloy is Al-2.5Cu-1.5Mg, aged 7 h at 200° C (from Vietz, J. T. and Polmear, I. J., *J. Inst. Metals*, **94**, 410, 1966)

upper temperature limit of stability of the GP zones for different compositions although its precise location can vary depending upon the concentration of excess vacancies. Solvus lines can also be determined for other metastable precipitates. The distribution of GP zone sizes with ageing time is shown schematically in Fig. 2.5. There is strong experimental support for the model proposed by Lorimer and Nicholson whereby GP zones formed below the GP zones solvus temperature can act as nuclei for the next stage in the ageing process, usually the intermediate precipitate, providing they have reached a critical size (d_{crit} in Fig. 2.5). On the basis of this model, alloys have been classified into three types:

(i) Alloys for which the quench-bath temperature and the ageing temperature are both above the GP zones solvus. Such alloys show little or no response to age hardening due to the difficulty of nucleating a finely dispersed precipitate. An example is the Al-Mg system in which quenching results in a very high level of supersaturation, but where hardening is absent in compositions containing less than 5 to 6% magnesium.

(ii) Alloys in which both the quench-bath and ageing temperatures are below the GP zones solvus, e.g. some Al-Mg-Si alloys.

(iii) Alloys in which the GP zones solvus lies between the quench-bath temperature and the ageing temperature. This situation is applicable in most age-hardenable aluminium alloys. Advantage may be taken of the nucleation of an intermediate precipitate from pre-existing GP zones of sizes above d_{crit} using two-stage or duplex ageing treatments. These are now applied to some alloys to improve certain properties and this is discussed in more detail in Section 3.5.5. It is particularly relevant with respect to the problem of stress-corrosion cracking in high-strength aluminium alloys.

2.1.3 Precipitate-free zones at grain boundaries

All alloys in which precipitation occurs have zones adjacent to grain boundaries which are depleted of precipitate and Fig. 2.6a shows comparatively wide zones in an aged, high-purity Al-Zn-Mg alloy. These precipitate-free zones (PFZs) are formed for two reasons. First, there is a narrow ($\sim$ 50 nm) region either side of a grain boundary which is depleted

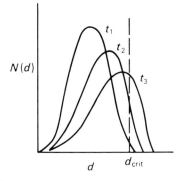

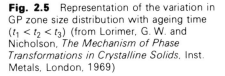

Fig. 2.5 Representation of the variation in GP zone size distribution with ageing time ($t_1 < t_2 < t_3$) (from Lorimer, G. W. and Nicholson, *The Mechanism of Phase Transformations in Crystalline Solids*, Inst. Metals, London, 1969)

of solute due to the ready diffusion of solute atoms into the boundary where relatively large particles of precipitate are subsequently formed. The remainder of a PFZ arises because of a depletion of vacancies to levels below that needed to assist with nucleation of precipitates at the particular ageing temperature. It has been proposed that the distribution of vacancies near a grain boundary can take the form shown schematically in Fig. 2.7 (curve A) and that a critical concentration C_1 is needed before nucleation of the precipitate can occur at temperature T_1. The width of the PFZ can be

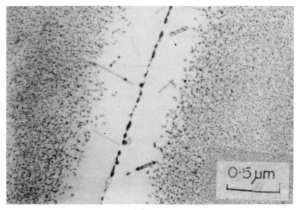

(a)

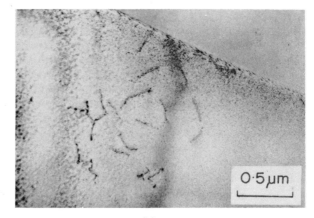

(b)

Fig. 2.6a   Wide PFZs in the alloy Al-4Zn-3Mg, aged 24 h at 150° C; b, effect of 0.3 % silver on PFZ width and precipitate distribution in Al-4Zn-3Mg, aged 24 h at 150° C (from Polmear, I. J., *J. Australian Inst. Met.*, **17**, 1, 1972)

altered by heat treatment conditions; the zones are narrower for higher solution treatment temperatures and faster quenching rates, both of which increase the excess vacancy content (e.g. curve B in Fig. 2.7), and for lower ageing temperatures. This latter effect has been attributed to a higher concentration of solute which means that smaller nuclei will be stable, thereby reducing the critical vacancy concentration required for nucleation to occur (C_2 in Fig. 2.7). However, the vacancy depleted part of a PFZ may be absent in some alloys aged at temperatures below the GP zones solvus as GP zones can form homogeneously without the need of vacancies.

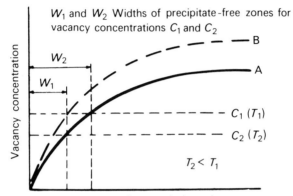

Fig. 2.7 Representation of profiles of vacancy concentration adjacent to a grain boundary in quenched alloys (from Taylor, J. L., *J. Inst. Metals,* **92**, 301, 1963–64)

2.1.4 Trace element effects

In common with other nucleation and growth processes, precipitation reactions may be greatly influenced by the presence of minor amounts or traces of certain elements. These changes can arise for a number of reasons including:

(i) preferential interaction with vacancies which reduces the rate of nucleation of GP zones

(ii) raising the GP zones solvus which alters the temperature ranges over which phases are stable

(iii) stimulating nucleation of an existing precipitate by reducing the interfacial energy between precipitate and matrix

(iv) promoting formation of a different precipitate.

Examples are the effects of minor additions of cadmium, indium, and tin in Al-Cu alloys (i) and (iii), silver in Al-Zn-Mg alloys (Fig. 2.6b) (ii), and silver in Al-Cu-Mg alloys (iv). Trace element effects can have important practical consequences in changing the properties and these are discussed when considering individual alloy systems.

2.1.5 Hardening mechanisms

Although early attempts to explain the hardening mechanisms in age-hardened alloys were limited by a lack of experimental data, two important concepts were postulated. One was that hardening, or the increased resistance of an alloy to deformation, was the result of interference to slip by particles precipitating on crystallographic planes. The other was that maximum hardening was associated with a critical particle size. Modern concepts of precipitation hardening are essentially the consideration of these two ideas in relation to dislocation theory, since the strength of an age-hardened alloy is controlled by the interaction of moving dislocations with precipitates.

Obstacles to the motion of dislocations in age-hardened alloys are the internal strains around precipitates, notably GP zones, and the actual precipitates themselves. With respect to the former, it can be shown that maximum impedence to the dislocation motion, i.e. maximum hardening, is to be expected when the spacing between particles is equal to the limiting radius of curvature of moving dislocation lines, i.e. about 50 atomic spacings or 10 nm. At this stage the dominant precipitate in most alloys is coherent GP zones, and high resolution transmission electron microscopy has revealed that these zones are, in fact, sheared by moving dislocations. Thus individual GP zones *per se* have only a small effect in impeding glide dislocations and the large increase in yield strength these zones may cause arises because of their high volume fraction.

Shearing of the zones increases the number of solute-solvent bonds across the slip planes in the manner depicted in Fig. 2.8 so that the process of clustering tends to be reversed. Additional work must be done by the applied stress in order for this to occur, the magnitude of which is controlled by factors such as relative atomic sizes of the atoms concerned and the difference in stacking fault energy between matrix and precipitate. This so-called chemical hardening makes an additional contribution to the overall strengthening of the alloy.

Once GP zones are cut, dislocations continue to pass through the particles on the active slip-planes and work hardening is comparatively small. Deformation tends to become localized on only a few active slip planes so that some intense bands develop. As will be discussed in Section 2.4, this microstructure may be deleterious with respect to other properties such as fatigue and stress-corrosion.

However, if precipitate particles are large and widely spaced they can be readily by-passed by moving dislocations which bow out between them and rejoin by a mechanism first proposed by Orowan (Fig. 2.9). Loops of dislocations are left around the particles. The yield strength of the alloy is low but the rate of work hardening is high, and plastic deformation tends to be spread more uniformly throughout the grains. This is the situation with

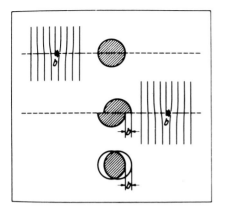

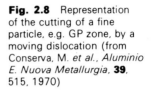

Fig. 2.8 Representation of the cutting of a fine particle, e.g. GP zone, by a moving dislocation (from Conserva, M. *et al.*, *Aluminio E. Nuova Metallurgia*, **39**, 515, 1970)

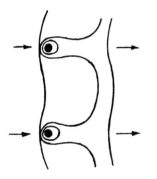

Fig. 2.9 Representation of a dislocation by-passing widely-spaced particles

over-aged alloys and the typical age-hardening curve in which strength increases then decreases with ageing time has been associated with a transition from shearing to by-passing precipitates, as shown schematically in Fig. 2.10.

The most interesting situation arises if precipitates are present which can resist shearing by dislocations and yet be too closely spaced to allow by-passing by dislocations. In such a case, the motion of dislocation lines would only be possible if sections can pass over or under individual particles by a process such as cross-slip. High levels of both strengthening and work hardening would then be expected. Normally such precipitates are too widely spaced for this to occur, but recent work involving duplex ageing treatments, below and above the GP zones solvus, has enabled the dispersion of certain intermediate precipitates in some commercial alloys to be refined with a consequent improvement in mechanical properties. A second possibility is to form duplex dispersions of precipitates, consisting of small, closely-spaced particles to raise the yield strength and larger

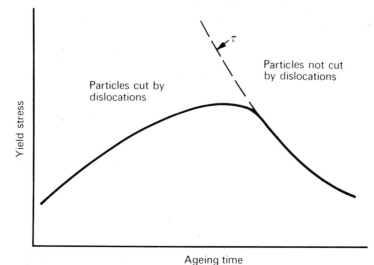

Fig. 2.10 Representation of the variation of yield stress with ageing time for a typical age-hardening alloy. τ is the shear stress needed to force dislocations between precipitate particles (from Kelly, A. and Nicholson R. B., *Progress in Materials Science*, **10**, 149, 1963)

particles which will cause increased rates of work hardening and distribute plastic deformation more uniformly.

2.2 Ageing processes

As mentioned earlier, several aluminium alloys display a marked response to age hardening. By suitable alloying and heat treatment, it is possible to increase the yield stress of high-purity aluminium by as much as 40 times. Details of the precipitation processes in alloy systems having commercial significance are shown in Table 2.2.

A partial phase diagram for the Al-Cu system is shown in Fig. 2.1 and the Al-Mg-Si system is represented by a pseudo-binary Al-Mg$_2$Si diagram in Fig. 2.11. Sections of phase diagrams for the ternary Al-Cu-Mg and Al-Zn-Mg systems are shown in Figs. 2.12 and 2.13. Most of the commercial alloys based on either system have additional alloying elements present which modify these diagrams, e.g. the section at 460° C for Al-Zn-Mg alloys containing 1.5 % copper which is shown in Fig. 2.14. This is the usual solution treatment temperature for alloys of this type and it should be noted that some quaternary compositions will not be single phase after such a treatment.

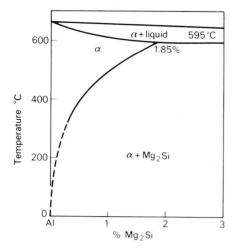

Fig. 2.11 Pseudo-binary phase diagram for Al-Mg$_2$Si

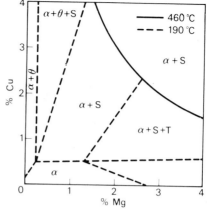

Fig. 2.12 Section of ternary Al-Cu-Mg phase diagram at 460° C and 190° C (estimated). θ = CuAl$_2$, S = Al$_2$CuMg, T = Al$_6$CuMg$_4$

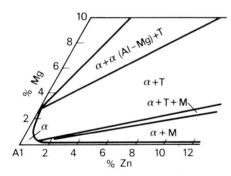

Fig. 2.13 Section of ternary Al-Zn-Mg phase diagrams at 200° C. M = MgZn$_2$. T = Al$_{32}$(Mg,Zn)$_{49}$

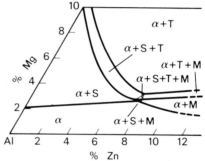

Fig. 2.14 Section of Al-Zn-Mg-Cu phase diagram (1.5 % Cu) at 460° C. S = Al$_2$CuMg, T = Al$_6$CuMg$_4$ + Al$_{32}$(Mg,Zn)$_{49}$, M = MgZn$_2$ + AlCuMg

2.3 Corrosion

2.3.1 Surface oxide film

Aluminium is an active metal which will oxidize readily under the influence of the high free energy of the reaction whenever the necessary conditions for oxidation prevail. Nevertheless, aluminium and its alloys are relatively

Table 2.2 Probable precipitation processes in aluminium alloys of commercial interest

Alloy	Precipitates	Remarks
Al-Cu	GP zones as thin plates on $\{100\}_\alpha$	Probably single layers of copper atoms on $\{100\}_\alpha$
	θ'' (formerly GP zones [2])	Coherent, probably two layers of copper atoms separated by three layers of aluminium atoms. May be nucleated at GP zones
	θ' tetragonal $CuAl_2$ $a = 0.404$ nm $c = 0.580$ nm	Semi-coherent plates nucleated at dislocations. Form on $\{100\}_\alpha$
	θ body-centred tetragonal $CuAl_2$ $a = 0.607$ nm $c = 0.487$ nm	Incoherent equilibrium phase. May nucleate at surface of θ'
Al-Mg ($> 5\%$)	Spherical GP zones	GP zones solvus below room temperature if $< 5\%$ Mg and close to room temperature in compositions between 5 and 10% Mg
	β' hexagonal $a = 1.002$ nm $c = 1.636$ nm	Probably semi-coherent. Nucleated on dislocations $(0001)_\beta//(001)_\alpha : [01\bar{1}0]_\beta//[110]_\alpha$
	β face-centred cubic Mg_5Al_8 (formerly Mg_2Al_3) $a = 2.824$ nm	Incoherent, equilibrium phase. Forms as plates or laths in grain boundaries and at a surface of β' particles in matrix $(111)_\beta//(001)_\alpha : [110]_\beta//[010]_\alpha$
Al-Si	Silicon diamond cubic $a = 0.542$ nm	Silicon forms directly from SSSS
Al-Cu-Mg	GP (Cu, Mg) zones as rods along $<100>_\alpha$	GP zones form very rapidly in most compositions aged at elevated temperatures
	S' orthorhombic Al_2CuMg $a = 0.404$ nm $b = 0.925$ nm $c = 0.718$ nm	Semi-coherent and nucleated at dislocations. Forms as laths in $\{210\}_\alpha$ along $<001>_\alpha$

Alloy	Phase / structure	Comments
	S orthorhombic Al$_2$CuMg a = 0.400 nm b = 0.923 nm c = 0.714 nm	Incoherent equilibrium phase, probably transforms from S'. Note that precipitates from the Al-Cu system can also form in compositions with high Cu:Mg ratios
Al-Mg-Si	GP zones as needles along $<100>_\alpha$	GP zones solvus occurs at temperatures that are normally higher than the ageing temperatures
	β' hexagonal Mg$_2$Si a = 0.705 nm c = 0.405 nm	Semi-coherent rods that probably form directly from GP zones. Lie along $<100>_\alpha$. $(001)_\beta //(100)_\alpha$: $[100]_\beta //[011]_\alpha$
	β face-centred cubic Mg$_2$Si a = 0.639 nm	Platelets on $\{100\}_\alpha$. May transform directly from β' $(100)_\beta //(100)_\alpha$: $[110]_\beta //[100]_\alpha$
Al-Zn-Mg	GP zones as spheres	Possibility of two types of zones
	η' (or M') hexagonal MgZn$_2$ a = 0.496 nm c = 0.868 nm	May form from GP zones in alloys with Zn:Mg > 3:1 $(0001)_\beta //(111)_\alpha$: $[11\bar{2}0]_\beta //[11\bar{2}]_\alpha$
	η (or M) hexagonal MgZn$_2$ a = 0.521 nm c = 0.860 nm	Forms at or from η', may have one of nine orientation relationships with matrix. Most common are: $(10\bar{1}0)_\eta //(001)_\alpha$: $(0001)_\eta //(110)_\alpha$: $(0001)_\eta //(111)_\alpha$: $(10\bar{1}0)_\eta //(110)_\alpha$
	T' hexagonal, probably Mg$_{32}$(Al, Zn)$_{49}$ a = 1.388 nm c = 2.752 nm	Semi-coherent. May form instead of η in alloys with high Mg:Zn ratios $(0001)_T //(111)_\alpha$: $(10\bar{1}1)_T //(11\bar{2})_\alpha$
	T cubic Mg$_{32}$(Al, Zn)$_{49}$ a = 1.416 nm	May form from η if ageing temperature >190°C, or from T' in alloys with high Mg:Zn ratios $(100)_T //(112)_\alpha$: $[001]_T //[1\bar{1}00]_\alpha$
Al-Mg-Li	δ' Al$_3$Li	Metastable ordered precipitate that is mainly responsible for strengthening
	Al$_2$MgLi	Incoherent equilibrium phase. Probably forms from δ'

stable in most environments due to the rapid formation of a natural oxide film of alumina on the surface that inhibits the bulk reaction predicted from thermodynamic data. Moreover, if the surface of aluminium is scratched sufficiently to remove the oxide film, a new film will quickly re-form in most environments. As a general rule, the protective film is stable in aqueous solutions of the pH range 4.5 to 8.5 whereas it is soluble in strong acids or alkalis leading to rapid attack of the aluminium. Exceptions are concentrated nitric acid, glacial acetic acid and ammonium hydroxide.

The oxide film formed on freshly rolled aluminium exposed to air is very thin and has been measured as 2.5 nm. It may continue to grow at a decreasing rate for several years to reach a thickness of some tens of nanometres. The rate of film growth becomes more rapid at higher temperatures and higher relative humidities, i.e. in water it is many times that occurring in dry air. In aqueous solutions, it has been suggested that the initial corrosion product is aluminium hydroxide which changes with time to become a hydrated aluminium oxide. The main difference between this film and that formed in air is that it is less adherent and so is far less protective.

Much thicker surface oxide films that give enhanced corrosion resistance to aluminium and its alloys can be produced by various chemical and electrochemical treatments. The natural film can be thickened some 500 times, say 1-2 μm, by immersion of components in certain hot acid or alkaline solutions. Although the films produced are mainly Al_2O_3, they also contain chemicals such as chromates which are collected from the bath to render them more corrosion resistant. A number of proprietary solutions are available and the films they produce are known generally as conversion coatings. Even thicker, e.g. 10–20 μm, surface films are produced by the more commonly used treatment known as anodizing. In this case the component is made the anode in an electrolyte, such as an aqueous solution containing 15 % sulphuric acid, which produces a porous Al_2O_3 film that is subsequently sealed, i.e. rendered non-porous, by boiling in water. Both conversion and anodic coatings can be dyed to give attractive colours and the latter process is widely applied to architectural products.

2.3.2 Contact with dissimilar metals

The electrode potential of aluminium with respect to other metals becomes particularly important when considering galvanic effects arising from dissimilar metal contact. Comparisons must be made by taking measurements in the same solution and Table 2.3 shows the electrode potentials with respect to the 0.1 M calomel electrode (Hg-HgCl$_2$, 0.1 M KCl) for various metals and alloys immersed in an aqueous solution of 1 M NaCl and 0.1 M H_2O_2. The value for aluminium is -0.85 V whereas aluminium alloys range from -0.69 V to -0.99 V. Magnesium which has an electrode potential of -1.73 V is more active than aluminium whereas mild steel is cathodic having a value of -0.58 V.

Table 2.3 Electrode potentials of various metals and alloys with respect to the 0.1 M calomel electrode in aqueous solutions of 53 g l^{-1} NaCl and 3 g l^{-1} H$_2$O$_2$ at 25°C (from *Metals Handbook*, Volume 1, American Society of Metals, Ohio, 1961)

Metal or alloy		Potential (V)
Magnesium		−1.73
Zinc		−1.10
Alclad 6061, Alclad 7075		−0.99
5456, 5083		−0.87
Aluminium (99.95%), 5052, 5086	aluminium alloys●	−0.85
3004, 1060, 5050		−0.84
1100, 3003, 6063, 6061, Alclad 2024		−0.83
2014-T4		−0.69
Cadmium		−0.82
Mild steel		−0.58
Lead		−0.55
Tin		−0.49
Copper		−0.20
Stainless steel (3xx series)		−0.09
Nickel		−0.07
Chromium		−0.49 to +0.18

● Compositions corresponding to the numbers are given in Tables 3.2 and 3.4

Table 2.3 suggests that sacrificial attack of aluminium and its alloys will occur when they are in contact with most other metals in a corrosive environment. However, it should be noted that electrode potentials serve only as a guide to the possibility of galvanic corrosion. The actual magnitude of the galvanic corrosion current is determined not only by the difference in electrode potentials between the particular dissimilar metals but also by the total electrical resistance, or polarization, of the galvanic circuit. Polarization itself is influenced by the nature of the metal/liquid interface and more particularly by the oxides formed on metal surfaces. For example, contact between aluminium and stainless steels usually results in less electrolytic attack than might be expected from the relatively large difference in the electrode potentials, whereas contact with copper causes severe galvanic corrosion of aluminium even though this difference is less.

2.3.3 Influence of alloying elements and impurities

Alloying elements may be present as solid solutions with aluminium, or as micro-constituents comprising the element itself, e.g. silicon, a compound between one or more elements and aluminium (e.g. Al$_2$CuMg) or as a compound between one or more elements (e.g. Mg$_2$Si). Any or all of the above conditions may exist in a commercial alloy. Table 2.4 gives values of the electrode potentials of some aluminium solid solutions and micro-constituents.

In general, a solid solution is the most corrosion resistant form in which an alloy may exist. Magnesium dissolved in aluminium renders it more anodic although dilute Al-Mg alloys retain a relatively high resistance to

Table 2.4 Electrode potentials of aluminium solid solutions and micro-constituents with respect to the 0.1 M calomel electrode in aqueous solutions of 53 g l^{-1} NaCl and 3 g l^{-1} H$_2$O$_2$ at 25°C (from *Metals Handbook*, Volume 1, American Society of Metals, Ohio, 1961)

Solid solution or micro-constituent	Potential (V)
Mg$_5$Al$_8$	−1.24
Al-Zn-Mg solid solution (4% MgZn$_2$)	−1.07
MgZn$_2$	−1.05
Al$_2$CuMg	−1.00
Al–5% Mg solid solution	−0.88
MnAl$_6$	−0.85
Aluminium (99.95%)	−0.85
Al-Mg-Si solid solution (1% Mg$_2$Si)	−0.83
Al-1% Si solid solution	−0.81
Al-2% Cu supersaturated solid solution	−0.75
Al-4% Cu supersaturated solid solution	−0.69
FeAl$_3$	−0.56
CuAl$_2$	−0.53
NiAl$_3$	−0.52
Si	−0.26

corrosion, particularly to sea water and alkaline solutions. Chromium, silicon and zinc in solid solution in aluminium have only minor effects on corrosion resistance although zinc does cause a significant increase in the electrode potential. As a result, Al-Zn alloys are used as clad coatings for certain aluminium alloys (see Section 3.1.3) and as galvanic anodes for the cathodic protection of steel structures in sea water. Copper reduces the corrosion resistance of aluminium more than any other alloying element and this arises mainly because of its presence in micro-constituents. However, it should be noted that when added in small amounts (0.05 to 0.2%), corrosion of aluminium and its alloys tends to become more general and pitting attack is reduced. Thus, although under corrosive conditions the overall weight loss is greater, perforation by pitting is retarded.

Micro-constituents are usually the source of most problems with electrochemical corrosion as they lead to non-uniform attack at specific areas of the alloy surface. Pitting and intergranular corrosion are examples of localized attack (Fig. 2.15) with exfoliation (layer) corrosion of components having a marked directionality of grain structure being an extreme example of this latter phenomenon (Fig. 2.16). In exfoliation corrosion, delamination of surface grains or layers occurs under forces exerted by the voluminous corrosion products.

Iron and silicon occur as impurities and form compounds most of which are cathodic with respect to aluminium. For example the compound FeAl$_3$ provides points at which the surface oxide film is weak, thereby promoting electrochemical attack. The rate of general corrosion of high-purity aluminium is much less than that of the commercial-purity grades which is attributed to the smaller size and number of these cathodic constituents throughout the grains. However, it should be noted that this may be a

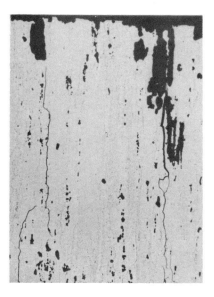

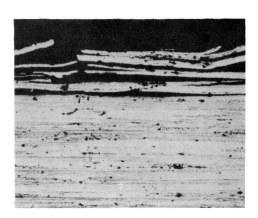

Fig. 2.15 Microsection of surface pits in a high-strength aluminium alloy. Note that intergranular stress-corrosion cracks are propagating from the base of these pits. ×100

Fig. 2.16 Microsection showing exfoliation (layer) corrosion of an aluminium alloy plate. ×100

disadvantage in some environments as attack of high-purity aluminium may be concentrated in grain boundaries. Nickel and titanium also form cathodic phases although nickel is present in very few alloys. Titanium, which forms $TiAl_3$, is commonly added to refine grain size (Section 3.1.1) but the amount is too small to have a significant effect on corrosion resistance. Manganese and aluminium form $MnAl_6$ which has almost the same electrode potential as aluminium and this compound is capable of dissolving iron which reduces the detrimental effect of this element. Magnesium in excess of that in solid solution in binary aluminium alloys tends to form the strongly anodic phase Mg_5Al_8 which precipitates in grain boundaries and promotes intercrystalline attack. However, magnesium and silicon, when together in the atomic ratio 2:1, form the phase Mg_2Si which also has a similiar electrode potential to aluminium.

2.3.4 Metallurgical and thermal treatments

Treatments that are carried out to change the shape and achieve a desired level of mechanical properties in aluminium alloys may also modify corrosion resistance, largely through their effects on both the quantity and the distribution of micro-constituents. In this regard, the complex changes associated with ageing or tempering treatments are on a fine scale and these are considered in Chapter 3. Both mechanical and thermal treatments can introduce residual stresses into components which may contribute to the phenomenon of stress-corrosion cracking and this is discussed in Section 2.4.4.

If one portion of an alloy surface receives a thermal treatment different from the remainder of the alloy, differences in potential between these regions can result. Welding processes provide an extreme example of such an effect and differences of up to 0.1 V may exist between the weld bead, heat affected zones and the remainder of the parent alloy.

Most wrought products do not undergo bulk recrystallization during subsequent heat treatment so that the elongated grain structure resulting from mechanical working is retained. Three principal directions are recognized: longitudinal, transverse (or long transverse) and short transverse, and these are represented in Fig. 2.17. This directionality of grain structure is significant in components when corrosion processes

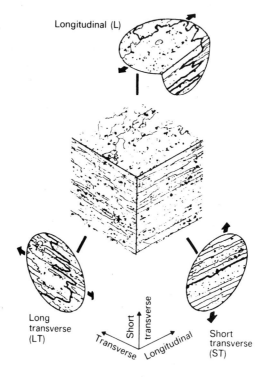

Fig. 2.17 The three principal directions with respect to the grain structure in a wrought aluminium alloy. Note the appearance of cracks that may form when stressing in these three directions (from Speidel, M. O. and Hyatt, M. V., *Advances in Corrosion Science and Technology*, Plenum Press, New York, 1972)

involve intercrystalline attack as has been illustrated by exfoliation corrosion. It is particularly important in regard to stress-corrosion cracking which is discussed in Section 2.4.4.

In certain products such as extrusions and die forgings, working is non-uniform and a mixture of unrecrystallized and recrystallized grain structures may form between which potential differences may exist. Large, recrystallized grains normally occur at the surface (see Fig. 3.6) and these are usually slightly cathodic with respect to the underlying, unrecrystallized grains. Preferential attack may occur if the relatively more anodic internal grains are partly exposed as may occur by machining.

2.4 Mechanical behaviour

The principal microstructural features that control the properties of aluminium alloys are:

(i) Coarse intermetallic compounds, usually in the range 0.5 to 10 μm, which form during ingot solidification or in subsequent processing, and which often contain the impurity elements iron and silicon. They include the virtually insoluble compounds $(Mn, Fe) Al_6$, $FeAl_3$, α-Al (Fe, Mn) Si, Al_7Cu_2Fe and the relatively soluble compounds $CuAl_2$, Mg_2Si and Al_2CuMg. As shown in Fig. 2.18, these particles tend to be aligned as stringers in fabricated products.

(ii) Smaller, submicron particles or dispersoids, 0.05 to 0.5 μm, which are intermetallic compounds containing the transition metals chromium, manganese or zirconium, or other high melting point elements, e.g.

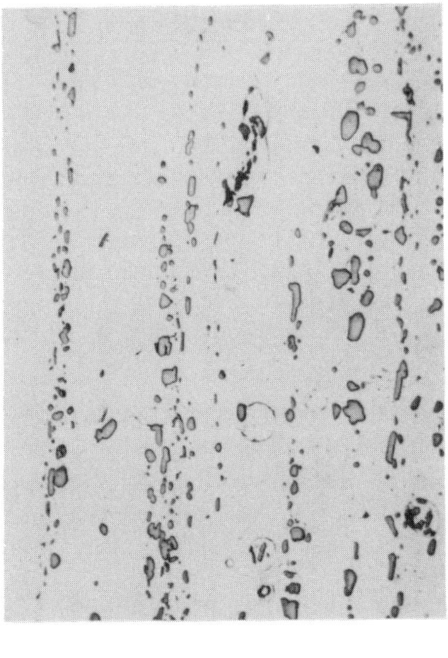

Fig. 2.18 Aligned stringers of coarse intermetallic compounds in a rolled aluminium alloy. ×250

$Al_{20}Cu_2Mn_3$, $Al_{12}Mg_2Cr$ and $ZrAl_3$. These particles serve primarily to retard recrystallization and grain growth in the alloys.

(iii) Fine precipitates, up to 0.01 μm, which form during age-hardening heat treatments and which promote strengthening.

(iv) Grain size and shape.

(v) Dislocation substructure, notably that formed during cold-working. Each of these features may be influenced by the various stages involved in the solidification and processing of wrought and cast alloys and these are discussed in detail in Chapters 3 and 4. It is now appropriate to consider the significance of the five features with respect to mechanical behaviour.

2.4.1 Tensile properties

Aluminium alloys may be divided into two groups depending upon whether or not they respond to precipitation hardening. The tensile properties of commercial wrought and cast compositions are considered in Chapters 3 and 4 respectively. For alloys that do respond to ageing treatments, it is the finely dispersed precipitates that have the dominant effect in raising yield and tensile strengths. For the other group, the dislocation substructure produced by cold-working in the case of wrought alloys and the grain size of cast alloys are of prime importance.

Coarse intermetallic compounds have relatively little effect on yield or tensile strength but they can cause a marked loss of ductility in both the cast and wrought products. The particles may crack at small plastic strains forming internal voids which, under the action of further plastic strain, may coalesce leading to premature fracture.

The fabrication of wrought products may cause highly directional grain flow. Moreover the resulting elongated grain structures are often deliberately retained by the addition of elements, e.g. manganese or chromium, that form submicron particles and prevent recrystallization during working and subsequent heat treatment. At the same time, the coarse intermetallics also become aligned to form stringers in the direction of metal flow (Fig. 2.18). These microstructural features are known as mechanical fibring and together with crystallographic texturing (Section 3.3.3) they cause anisotropy in tensile and other properties. Accordingly measurements are often made in the three principal directions shown in Fig. 2.17. Tensile properties, notably ductility, are greatest in the longitudinal direction and least in the short transverse direction in which stressing is normal to the stringers of intermetallics, e.g. Table 2.5.

2.4.2 Toughness

Early work on the higher strength aluminium alloys was directed primarily at maximizing tensile properties in materials for aircraft construction. More recently, the emphasis in alloy development has shifted away from tensile strength as an over-riding consideration and more attention is being given to the behaviour of alloys under the variety of conditions en-

Table 2.5 Variation in tensile properties with direction in 76 mm thick aluminium alloy plates (from Forsyth, P. J. E. and Stubbington, A., *Metals Technology*, **2**, 158, 1975)

Alloy direction	0.2% proof stress (MPa)	Tensile strength (MPa)	Elongation (%)
Al-Zn-Mg-Cu (7075)			
Longitudinal (L)	523	570	15.5
Long transverse (LT)	482	552	12.0
Short transverse (ST)	445	527	7.5
ST/L ratio	0.85	0.93	0.48
Al-Cu-Mg (2014)			
Longitudinal (L)	441	477	14.0
Long transverse (LT)	423	471	10.5
Short transverse (ST)	404	449	4.0
ST/L ratio	0.91	0.94	0.29

countered in service. Tensile strength controls resistance to failure by mechanical overload but, in the presence of cracks and other flaws, it is the toughness (and more particularly the fracture toughness) of the alloy that becomes the most important parameter.

In common with other metallic materials, the toughness of aluminium alloys decreases as the general level of strength is raised by alloying and heat treatment. Minimum fracture toughness requirements become more stringent and, in the high-strength alloys, it is necessary to place a ceiling on the level of yield strength that can be safely employed by the designer.

Crack extension in commercial aluminium alloys proceeds by the ductile, fibrous mode involving the growth and coalescence of voids nucleated by cracking or by decohesion at the interface between second phase particles and the matrix (Fig. 2.19). Consequently, the important metallurgical factors are:

(i) the distribution of the particles that crack

(ii) the resistance of the particles and their interfaces with the matrix to cleavage and decohesion

(iii) the local strain concentrations which accelerate coalescence of the voids

(iv) the grain size when coalescence involves grain boundaries.

The major step in the development of aluminium alloys with greatly improved fracture toughness has come from the control of the levels of the impurity elements iron and silicon. This effect is shown in Fig. 2.20 for alloys based on Al-Cu-Mg system and it can be seen that plane strain fracture toughness values may be doubled by maintaining the combined levels of these elements below 0.5% as compared with similar alloys in which this value exceeds 1.0%. As a consequence of this, a range of high-toughness versions of older alloy compositions are now in commercial use in which the levels of impurities have been reduced (see Table 3.4).

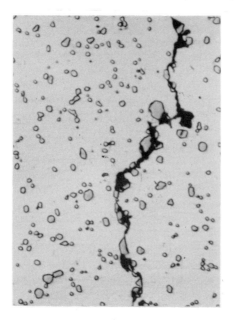

Fig. 2.19 Microstructure of an aluminium casting alloy in which the crack path has been influenced by coarse silicon particles (courtesy R. W. Coade). ×700

The role of the submicron dispersoids with respect to toughness is more complex as they have both good and bad effects. To the extent that they suppress recrystallization and limit grain growth they are beneficial. The effect of these factors on a range of high-strength sheet alloys based on the Al-Zn-Mg-Cu system is shown in Fig. 2.21. Fine, unrecrystallized grains favour a high energy absorbing, transcrystalline mode of fracture. On the other hand, such particles also nucleate microvoids by decohesion at the interface with the matrix which may lead to the formation of sheets of voids between the larger voids that are associated with the coarse intermetallic compounds. In this regard, the effects do vary with different transition metals and there is evidence to suggest that alloys containing zirconium to control grain shape are more resistant to fracture than those to which chromium or manganese have been added for this purpose. This is attributed to the fact that zirconium forms relatively small particles of the compound $ZrAl_3$ which are around 20 nm in diameter.

The fine precipitates developed by age hardening are also thought to have at least two effects with regard to the toughness of aluminium alloys. To the extent that they reduce deformation, toughness is enhanced and it has been observed that, for equal dispersions of particles, an alloy with a higher yield stress has greater toughness. At the same time, these fine particles tend to cause localization of slip during plastic deformation, particularly under plane strain conditions, leading to development of pockets of slip or so-called superbands ahead of an advancing crack. Strain is concentrated within these bands and may cause premature cracking at the sites of intermetallic compounds ahead of an advancing crack. This

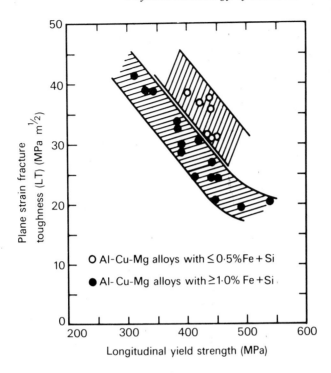

Fig. 2.20 Plane strain fracture toughness of commercial Al-Cu-Mg sheet alloys with differing levels of iron and silicon (from Speidel, M. O., *Proceedings of 6th International Conference on Light Metals*, Leoben, Austria, Aluminium-Verlag, Düsseldorf, 1975)

effect is usually greatest for alloys aged to peak hardness and the overall result is a net loss of toughness at the highest strength levels.

The fact that toughness does not follow a simple inverse relationship to the strength of age-hardened aluminium alloys is shown by comparing alloys in the under- and over-aged conditions. Toughness is greatest in under-aged conditions and, although some recovery from the minimum value that coincides with peak strength occurs on over-ageing, the presence of coarser precipitates does partly offset the gains arising from a reduction in yield strength. This effect varies with different alloys and is particularly marked with alloys based on the Al-Cu-Mg system due to the detrimental influence of the relatively coarse S′, or S, phase (Al_2CuMg) precipitates that form on over-ageing. There is also a greater tendency to fracture along grain boundaries as ageing proceeds.

2.4.3 Fatigue

It is well known that, contrary to steels, the increases that have been

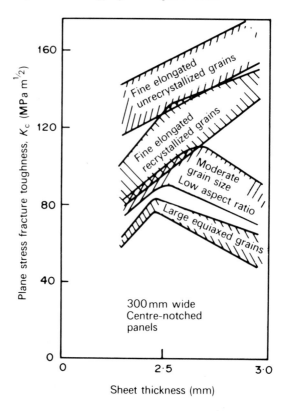

Fig. 2.21   Effect of recrystallization and grain size and shape of various alloys based on the Al-Zn-Mg-Cu system (from Thompson, D .S., *Met. Trans.*, **6A**, 671, 1975)

achieved in the tensile strength of most non-ferrous alloys have not been accompanied by proportionate improvements in fatigue properties. This feature is illustrated in Fig. 2.22 which shows relationships between fatigue endurance limit (5 x 10^8 cycles) and tensile strength for different alloys. It should also be noted that the so-called fatigue ratios are lowest for age-hardened aluminium alloys and, as a general rule, the more an alloy is dependent upon precipitation hardening for its total strength, the lower this ratio becomes.

Detailed studies of the processes of fatigue in metals and alloys have shown that the initiation of cracks normally occurs at the surface. It is here that strain becomes localized due to the presence of pre-existing stress concentrations such as mechanical notches or corrosion pits, coarse (persistent) slip bands in which minute extrusions and intrusions may form, or at relatively soft zones such as the precipitate-free regions adjacent to grain boundaries.

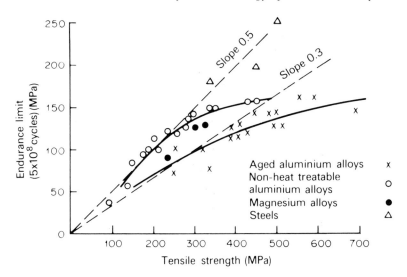

Fig. 2.22 Fatigue ratios (endurance limit:tensile strength) for aluminium alloys and other materials (from Varley, P. C., *The Technology of Aluminium and its Alloys*, Newnes-Butterworths, London, 1970)

The disappointing fatigue properties of age-hardened aluminium alloys are also attributed to an additional factor which is the metastable nature of the metallurgical structure under conditions of cyclic stressing. Localization of strain is particularly harmful because the precipitate may be removed from certain slip bands which causes softening there and leads to a further concentration of stress so that the whole process of cracking is accelerated. This effect is shown in an exaggerated manner in a recrystallized, high-purity alloy in Fig. 2.23. It has been proposed that removal of the precipitate occurs either by over-ageing or re-solution, the latter now being considered to apply in most cases. One suggestion is that the particles in the slip bands are cut by moving dislocations, and re-solution occurs when they become smaller than the critical size for thermodynamic stability.

The fatigue behaviour of age-hardened aluminium alloys should, therefore, be improved if fatigue deformation could be dispersed more uniformly. Factors which prevent the formation of coarse slip bands should assist in this regard. Thus it is to be expected that commercial-purity alloys should perform better than equivalent high-purity compositions because the presence of inclusions and intermetallic compounds would tend to disperse slip. This effect has been confirmed for an Al-Cu-Mg alloy, and fatigue curves for commercial-purity and high-purity compositions are shown in Fig. 2.24. Here the superior fatigue behaviour of the former alloy arises because slip is more uniformly dispersed by submicron dispersoids such as $MnAl_6$. Thermomechanical processing whereby plastic deformation before, or during, the ageing treatment increases the dislocation

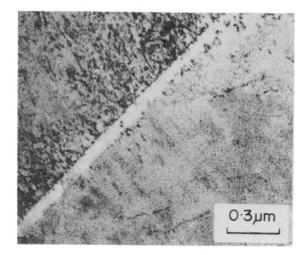

Fig. 2.23 Transmission electron micrograph showing precipitate depletion in a persistent slip band formed by fatigue stressing a high-purity Al-Zn-Mg alloy (courtesy A. Stubbington, copyright HMSO)

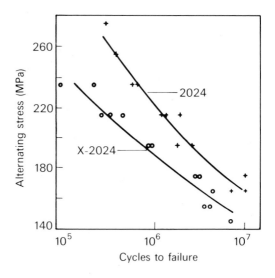

Fig. 2.24 Effect of reducing the concentration of submicron particles in an Al-Cu-Mg alloy. X2024 is a high-purity version of the commercial alloy 2024 (from Lütjering, G. *et al., Proceedings of Third International Conference on Strength of Metals and Alloys*, Inst. Metals, London)

density has also been found to improve the fatigue performance of certain alloys although this effect arises in part from an increase in tensile properties caused by such a treatment (Fig. 2.25). It should be noted, however, that the promising results mentioned above were obtained for smooth specimens. The improved fatigue behaviour has not been sustained for severely notched conditions and it seems that the resultant stress concentrations over-ride the more subtle microstructural effects that have been described.

Alloys which are aged at higher temperatures, and thus form relatively more stable precipitates, might also be expected to show better fatigue properties and this trend is observed. For example, the fatigue performance of the alloys based on the Al-Cu-Mg system is generally better than that of Al-Zn-Mg-Cu alloys, although this effect is again greatly reduced for notched conditions.

The fact that microstructure can have a greater influence upon the fatigue properties of aluminium alloys than the level of tensile properties has been demonstrated for an Al-Mg alloy containing a small addition of silver. It is well known that binary Al-Mg alloys such as Al-5Mg, in which the magnesium is present in solid solution, display a relatively high level of fatigue strength. The same applies for an Al-5Mg-0.5Ag alloy in the as-

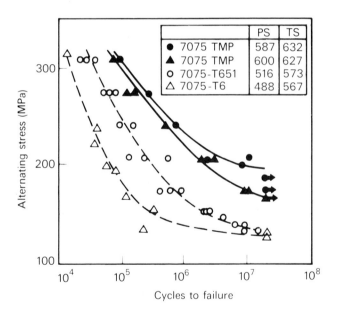

	PS	TS
● 7075 TMP	587	632
▲ 7075 TMP	600	627
○ 7075-T651	516	573
△ 7075-T6	488	567

Fig. 2.25 Effect of thermomechanical processing (TMP) on the unnotched fatigue properties of the commercial Al-Zn-Mg-Cu alloy 7075. PS = proof stress (MPa), TS = tensile strength (MPa) (from Ostermann, F. G., *Met. Trans.*, **2A**, 2897, 1971)

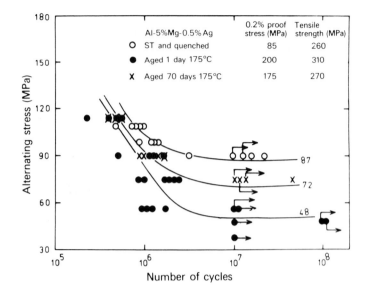

Fig. 2.26　Fatigue (*S/N*) curves for the alloy Al-5Mg-0.5Ag in different conditions (from Boyapati, K. and Polmear, I. J., *Fatigue of Engineering Materials and Structures*, **2**, 23, 1979)

quenched condition and Fig. 2.26 shows that the endurance limit after 10^8 cycles is ± 87 MPa which approximately equals the 0.2% proof stress. This result is attributed to the interaction of magnesium atoms with dislocations which minimizes formation of coarse slip bands during fatigue. The silver-containing alloy responds to age hardening at elevated temperatures due to the formation of a finely dispersed precipitate, and the 0.2% proof stress may be raised to 200 MPa after ageing for one day at 175° C. However, the endurance limit for 10^8 cycles is actually decreased to ± 48 MPa due to the localization of strain in a limited number of coarse slip bands (Fig. 2.27a).

Continued ageing of the alloy at 175°C causes only slight softening (0.2% proof stress 175 MPa after 70 days) although large particles of a second precipitate are formed (Fig. 2.27b) which have the effect of dispersing dislocations generated by cyclic stressing. As a result, fatigue properties are improved and the endurance limit for 10^8 cycles was raised to ± 72 MPa (Fig. 2.26). These particles serve the same role as the submicron particles in the commercial alloy 2024 (Fig. 2.24) but they have formed by a precipitation process. This again suggests the desirability of having a duplex precipitate structure; fine particles to give a high level of tensile properties, and coarse particles to improve fatigue strength.

2.4.4 Stress-corrosion cracking

Stress-corrosion cracking (SCC) may be defined as a phenomenon which results in brittle failure in alloys, normally considered ductile, when they

(a)   (b)

Fig. 2.27a   Coarse slip bands containing a high density of dislocations. Alloy Al-5Mg-0.5Ag aged one day at 175° C and tested at a stress of ±75 MPa for 1.4 ×10⁶ cycles; b, large particles of a second precipitate, that have formed in the alloy aged 70 days at 175° C, which have dispersed dislocations (see arrow) generated by fatigue stressing for 10⁷ cycles at a stress of ±75 MPa

are exposed to the simultaneous action of surface tensile stress and a corrosive environment, neither of which when operating separately could cause major damage. The level of stress needed for crack initiation and growth can be well below that causing yielding and the specific corrosive environments, e.g. water vapour, can be very mild. High-strength, wrought aluminium alloys have accounted for many failures by SCC in service and particular attention has been given to the development of modified compositions and heat treatment practices as ways of minimizing this problem. These aspects are discussed in more detail in Section 3.5.5.

SCC in aluminium alloys always takes place along grain boundaries and maximum susceptibility occurs in the recrystallized condition. For this reason compositions, working procedures and heat treatment temperatures for wrought alloys are normally adjusted to prevent recrystallization. It should be noted, however, that the resistance of a particular wrought alloy to SCC will now vary depending upon the direction of stressing with respect to the elongated grain structure. Maximum susceptibility occurs if stressing is normal to the grain direction, i.e. in the short transverse direction of components because the crack path along grain boundaries is so clearly defined (Fig. 2.17).

Considerable effort has been directed towards understanding the mechanism of SCC in aluminium alloys and significance has been attached to the following microstructural features:

(i) Precipitate-free zones adjacent to grain boundaries (Fig. 2.6) – in a corrosive medium, it is considered that either these zones or the grain boundaries will be anodic with respect to the grain centres. Moreover, strain is likely to be concentrated in the zones because they are relatively soft.

(ii) Nature of the matrix precipitate – maximum susceptibility to cracking occurs in alloys when GP zones are present. In this condition, deformation tends to be concentrated in discrete slip bands similar in appearance to those shown in Fig. 2.27a. It is considered that stress is generated where these bands impinge upon grain boundaries which can contribute to intercrystalline cracking under stress-corrosion conditions.

(iii) Dispersion of precipitate particles in grain boundaries – in some aged aluminium alloys, it has been shown that SCC occurs more rapidly when particles in grain boundaries are closely spaced.

(iv) Solute concentrations in the region of grain boundaries – differences in solute levels that arise during ageing are thought to modify local electrochemical potentials. Moreover, it has been observed that a higher magnesium content develops in these regions. This results in an adjacent oxide layer with an increased MgO content which, in turn, is a less effective barrier against environmental influences.

(v) Hydrogen embrittlement that may occur due to the rapid diffusion of hydrogen along grain boundaries.

(vi) Chemisorption of atom species at the surface of cracks which may lower the cohesive strength of the interatomic bonds in the region ahead of an advancing crack.

Recent experimental work has shown that stress-corrosion cracking at grain boundaries occurs in a brittle and discontinuous manner and there is clear evidence that hydrogen diffuses there, even in the absence of stress (e.g. Fig. 2.28). It thus seems that the presence of hydrogen does play a vital part in SCC due to one or both of mechanisms (v) and (vi). However, the overall process of SCC is complex and it seems probable that one or more of the other microstructural factors will also be involved. The relative importance of each of these factors may depend upon the particular combination of alloy and environment.

2.4.5 Corrosion-fatigue

Under conditions of simultaneous corrosion and cyclic stressing (corrosion-fatigue), the reduction in strength is greater than the additive effects if each is considered either separately or alternately. Although it is often possible to provide adequate protection for metallic parts which are stressed under static conditions, most surface films (including naturally protective oxides) can be more easily broken or disrupted under cyclic loading.

In general, the reduction in a fatigue strength of a material in a particular

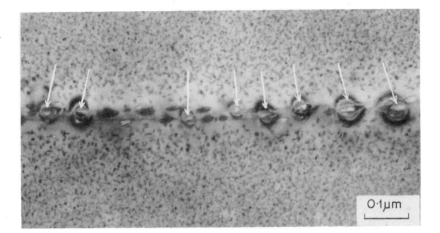

Fig. 2.28 Transmission electron micrograph showing hydrogen bubble development at precipitate particles in a grain boundary of a thin foil of an artificially aged Al-Zn-Mg alloy exposed to laboratory air for three months (from Scamans, G. M., *J. Materials Science*, **13**, 27, 1978)

corrosive medium will be related to the corrosion resistance of the material in that medium. Under conditions of corrosion-fatigue all types of aluminium alloys exhibit about the same percentage reduction in strength when compared with their fatigue strength in air. For example, under freshwater conditions the fatigue strength at 10^8 cycles is about 60 % that in air and in NaCl solutions it is normally between 25 to 35 % of that in air. Another general observation is that the corrosion-fatigue strength of a particular aluminium alloy appears to be virtually independent of its metallurgical condition.

2.4.6 Creep

Creep fracture, even in pure metals, normally occurs by the initiation of cracks in grain boundaries. The susceptibility of this region to cracking in age-hardened aluminium alloys is enhanced because the grains are harder and less willing to accommodate deformation than the relatively softer precipitate-free zones adjacent to the boundaries (Fig. 2.6). Moreover the strength of the grain boundaries may be modified by the presence there of precipitate particles.

Precipitation-hardened alloys are normally aged at one or two temperatures which allow peak properties to be realized in a relatively short time. Continued exposure to these temperatures normally leads to rapid over-ageing and softening and it follows that service temperatures must be well below the final ageing temperature if a loss of strength due to over-ageing is to be minimized. For example, the alloy selected for the structure and skin of the supersonic Concorde aircraft, which is normally required to operate in service at 100 to 110°C, is aged initially at 190°C.

Creep resistance in aluminium alloys is promoted by the presence of submicron intermetallic compounds such as $FeNiAl_9$ or other fine particles that are stable at the required service temperatures (normally below 200°C). Fine alumina particles which can be introduced by powder metallurgy methods can also serve the same purpose. Each type of particle provides some dispersion strengthening and also has the effect of reducing grain boundary migration.

Further reading

Kelly, A. and Nicholson, R. B., Precipitation hardening, *Progress in Materials Science*, Volume 10, Pergamon, London, 1963; p. 149

Van Horn, K. (ed.), *Aluminum*, Volume 1, American Society for Metals, Cleveland, 1967

Lorimer, G. W., In: *Precipitation Processes in Solids*, ed. K. C. Russell and H. I. Aaronson, Met. Soc., AIME, New York, 1978; Chapter 3

Polmear, I. J., In: *Physics of Materials*, ed. D. W. Borland *et al.*, CSIRO – University of Melbourne, 1979; p. 209

Mondolfo, L. F., *Aluminium Alloys: Structure and Properties*, Butterworths, London, 1976

Godard, H. P. *et al.*, *Corrosion of Light Metals*, Wiley, New York, 1967

Conservo, M. *et al.*, Ageing mechanisms and hardening structures in age-hardenable aluminium alloys, *Aluminio E. Nuova Metallurgia*, **39**, 515, 1970

Polmear, I. J., Nucleation from supersaturated solid solutions, *J. Australian Inst. Metals*, **11**, 246, 1966

Thompson, D. S., Metallurgical factors affecting high strength aluminium alloy production, *Met. Trans.*, **6A**, 671, 1975

Speidel, M. O., *Proc. 6th International Conference on Light Metals*, Leoben, Austria, Aluminium-Verlag, Düsseldorf, 67, 1975

Hahn, G. T. and Rosenfield, A. R., Metallurgical factors affecting fracture toughness of aluminium alloys, *Met. Trans.*, **6A**, 653, 1975

Garrett, G. G. and Knott, J. F., The influence of compositional and microstructural variations on the mechanism of static fracture in aluminium alloys, *Met. Trans.*, **9A**, 1187, 1978

Scamans, G. M., Pre-exposure embrittlement and stress corrosion failure in Al-Zn-Mg alloys, *Corros. Sci.*, **16**, 443, 1976

3
Wrought aluminium alloys

As a general average, about 85% of aluminium is used for wrought products e.g. rolled plate (> 6 mm thickness), sheet (0.15 mm to 6 mm), foil (< 0.15 mm), extrusions, tube, rod, bar and wire. These are produced from cast ingots, the structures of which are greatly changed by the various working operations and thermal treatments. Each class of alloys behaves differently, with composition and structure dictating the working characteristics and subsequent properties that are developed. Before considering individual alloys, it is necessary to examine how they are produced and heat treated.

3.1 Production of wrought alloys

3.1.1 Melting and casting

Ingots are prepared for subsequent mechanical working by first melting virgin aluminium, scrap and the alloying additions, usually in the form of a concentrated hardener or master alloy, in a suitable furnace. A fuel-fired reverberatory type is most commonly used. The main essentials in promoting ingot quality are a thorough mixing of the constituents together with effective fluxing and degassing of the melt before casting in order to remove dross, oxides, gases and other non-metallic impurities. Hydrogen is the only gas with measurable solubility in aluminium, the respective equilibrium solubilities in the liquid and solid states at the melting temperature and a pressure of one atmosphere being 0.68 and 0.036 cm^3 per 100 g of metal – a difference of 19 times. Atomic size factors require that hydrogen enters solution in the atomic form and the gas is derived from the surface reaction of aluminium with water vapour:

$$2Al + 3H_2O \rightarrow Al_2O_3 + 3H_2$$

The standard free energy change for this reaction is very high (equilibrium constant = 8×10^{40} at $725°C$) so that, for practical purposes, all traces of water vapour contacting the metal are converted to hydrogen. The main sources of water vapour are the furnace atmosphere in contact with the molten metal, hydrated surface contaminants and residual moisture in launders and moulds.

During solidification, excess hydrogen is rejected from solution and it recombines as molecular gas which may be entrapped in the solid structure leading to porosity, notably in interdendritic regions. In order to obtain an ingot that is free from gas porosity, it has been found necessary to reduce the hydrogen content of the molten metal to less than 0.15 cm³ per 100 g. Although hydrogen can escape from molten aluminium by evaporation from the surface, the process is slow and it is common practice to purge by bubbling an insoluble gas through the melt in the reverberatory furnace prior to pouring. Partial pressure requirements cause hydrogen to diffuse to these bubbles and be carried out of the metal. According to circumstances either nitrogen, argon, chlorine, mixtures of these gases or a solid chlorinated hydrocarbon is used. Chlorine is normally present because it serves the important additional function of increasing the surface tension between inclusions and the melt so that they tend to rise to the top and can be skimmed off. However, the use of chlorine does present environmental problems and considerable interest has been shown in the continuous fumeless in-line degassing (FILD) process developed by the British Aluminium Company. In this process, the molten aluminium from the melting furnace enters a vessel through a liquid flux cover (KCl + NaCl eutectic with small amounts of CaF₂). It is then degassed with nitrogen under conditions which do not increase the inclusion content and passes through a bed of balls of Al₂O₃ which become coated with flux and serve to further reduce the content of non-metallic inclusions in the aluminium. This process eliminates the capital and operating costs of fume treatment, as well as the need to hold the aluminium in a furnace whilst degassing is carried out. Loss of metal as dross is also reduced.

The production of a uniform ingot structure is desirable and this is promoted by direct-chill (DC), semi-continuous processes. Most commonly, ingots are cast by the vertical process in which the molten alloy is poured into one or more fixed water-cooled moulds having retractable bases (Fig. 3.1). The process of solidification is accomplished in two stages:

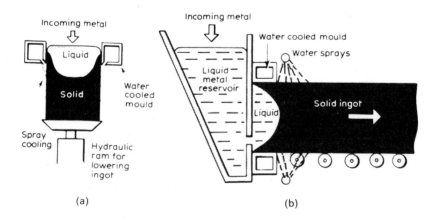

(a)

(b)

Fig. 3.1 Direct-chill casting processes: a, vertical; b, horizontal (from Lewis, D., *New Scientist*, **14**, 325, 1970)

formation of solid metal at the chilled mould wall and solidification of the remainder of the billet cross-section by the removal of heat by sub-mould, spray cooling. Ingots may be rectangular for subsequent rolling or forging, or round for extrusion and weigh several tonnes. Similar sections but in general of smaller dimensions may be produced in the more recently designed horizontal (Ugine) arrangement shown in Fig. 3.1, although control of microstructure, notably grain size, is more difficult in this process.

A novel alternative method of producing large ingots has been developed in the Soviet Union. This involves casting in a strong electromagnetic field which supports the molten metal and allows solidification to occur without contact with a mould. A much smoother surface is obtained on the ingots which obviates the need for scalping or machining prior to extrusion or rolling.

The concept of the moving mould has revolutionized the casting of some lower strength aluminium alloys as it is now possible to produce continuous shapes in sizes close to the requirements of the final wrought products, thereby reducing the investment in the heavy equipment required for subsequent mechanical working. Methods for casting bar and sheet are shown in Fig. 3.2 and each offers considerable economic advantages over earlier practices involving extrusion or rolling of large ingots to small section sizes.

Rate of solidification has an important influence on quality of an ingot, faster rates are desirable as they lead to finer dendrite arm spacings which, in turn, reduce microsegregation in the interdendritic spaces. The sizes of intermetallic compounds are also reduced and the final grain size is smaller and more uniform. Iron tends to be the most troublesome impurity as it may cause formation of relatively insoluble plates of the compound

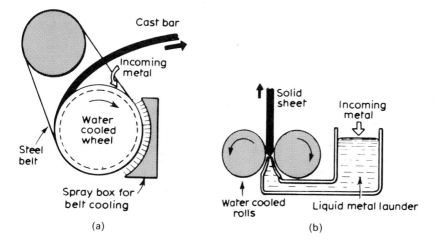

Fig. 3.2 Processes for continuously casting bar and sheet in moving moulds: a, Properzi process; b, Hunter process (from Lewis, D., *New Scientist*, **14**, 325, 1970)

β-AlFeSi which are brittle and may crack during subsequent working of an ingot. Fast rates of solidification also are beneficial in refining the size of these plates and, in some alloys, manganese in excess of 0.3 % is added to promote formation of α-AlFeSi which has a more desirable scriptic morphology.

Because fine grain size is so desirable, it is usual to add small amounts of master alloys of Al-Ti or Al-Ti-B (Ti:B in the range 5:1 to 100:1) to the melt in order to promote a further refinement. Two theories have been proposed to explain the mechanism of grain refinement, both of which have been keenly debated for many years. One is based on the heterogeneous nucleation of aluminium alloy grains at insoluble particles of TiC or TiB_2, in the melt, which have crystal structures similar to aluminium. The TiC forms at high temperatures by reaction between titanium and the carbon impurities that are always present. The second theory attributes grain refinement to the following peritectic reaction involving particles of the compound $TiAl_3$:

Liquid aluminium $+ TiAl_3 \rightarrow \alpha$-alloy solid solution

The newly formed solid solution will coat the particles of $TiAl_3$ giving nuclei from which grains may grow. Under certain conditions, these particles adopt a petal-like form (Fig. 3.3) and the crystal facets at each petal tip can be the point of origin of a separate α-aluminium dendrite without the need of the intermediate stage of forming a peritectic envelope. Multiple nucleation of as many as twelve new grains may occur on each particle, although the average is nearer to eight.

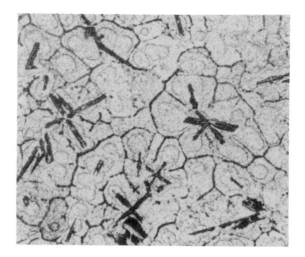

Fig. 3.3   Petal-like form of $TiAl_3$ particles in a matrix of α-aluminium solid solution. Note that several grains have been nucleated at each particle (from St John, D. H. and Hogan, L. M., *J. Australian Inst. Metals*, **22**, 160, 1977)

In practice, both of these mechanisms can apply, nucleation of single α-aluminium grains taking place on TiC particles if the titanium content is below that needed to initiate the peritectic reaction, and multiple nucleation of grains occurring by this reaction for higher titanium contents. In the absence of boron, the critical level of titanium is $\sim 0.10\%$ but this amount may be as little as 0.01 % if boron is present. Cornish has proposed that this change arises from the effect of boron in lowering the titanium content of the peritectic (Fig. 3.4). In addition, X-ray microanalysis of TiAl$_3$ particles has indicated that they may, themselves, nucleate on insoluble TiB$_2$ or (TiAl) B$_2$ particles if boron is present.

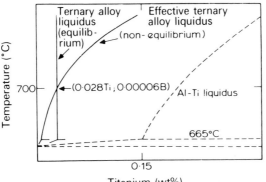

Fig. 3.4 Representation of the effective Al-Ti liquidus line for Al-0.05Ti-0.01B, initially at 700°C, under equilibrium and non-equilibrium conditions. Broken lines represent the binary Al-Ti equilibrium diagram (from Cornish, A. J., *Metal Science*, **9**, 447, 1975)

Removal of oxide skins and other particulate matter is effected partly during the degassing step, where this involves chlorine, and partly by screening the metal through glasscloth or by passing it through beds of refractory granules or rigid ceramic filters. Where granular beds are employed, the cleaning and degassing steps may usefully be combined into a single in-line operation. Apart from the use of fluid-type fluxes in the FILD process already mentioned, the use of salt fluxes is normally confined to furnace skimming operations for which a suitable composition would be NaCl 40 %, NaF 25 %, CaF$_2$ 6 %, AlF$_3$ 6 %, Na$_2$SO$_4$ 23 %.

Most ingots are inspected to ensure freedom from defects and cracks, the degree of testing and the acceptance standards depending upon the final usage of the material. The method commonly used is ultrasonic flaw detection.

3.1.2 Homogenization of ingots

Before DC ingots are fabricated into semi-finished forms, it is necessary to homogenize at a high temperature to reduce segregation and to remove non-equilibrium, low melting point eutectics that may cause cracking during subsequent working. In this regard, the homogenization time at a particular temperature has been found to be inversely related to the square of the dendrite arm spacing in the ingot. Homogenization is particularly important for the higher strength alloys as it also serves to precipitate and redistribute the submicron intermetallic compounds of the transition metals such as $MnAl_6$, $Al_{12}Mg_2Cr$ and $ZrAl_3$ which were discussed in Section 2.4. These transition metals may supersaturate in the aluminium during the relatively rapid cooling of DC ingots and it is then necessary to promote precipitation of uniform dispersions of the compounds in order to control grain structure. In addition, it is now realized that they may have a marked influence on various mechanical properties through their effects on both the subsequent response to ageing treatments and on the dislocation substructures formed by deformation.

Regulation of these various functions requires a careful choice of conditions for homogenizing ingots of different alloys. When precipitation of these compounds is involved, both time and temperature are significant and the rate of heating to the homogenization temperature is of crucial importance. Relatively slow rates, e.g. 75° C h^{-1}, are necessary to promote nucleation and growth of a fine and uniform dispersion of the compounds. It has been found that the compounds are actually nucleated heterogeneously at the surfaces of precipitate particles which form and grow to relatively large sizes during the slow heating cycle. Once formed, the submicron compounds remain stable at the homogenization temperature, whereas the precipitates are redissolved. This effect is illustrated for an Al-Zn-Mg alloy in Fig. 3.5. In this case, the precipitate η-$(MgZn_2)$ has provided an interface for the subsequent heterogeneous nucleation of $MnAl_6$ in the ingot that was heated at the slower rate.

Further modification to the size of these submicron particles is possible through control of the rates of cooling of ingots after homogenization is completed.

3.1.3 Fabrication of DC ingots

The next stage in the production of wrought alloys is the conversion of the ingot into semi-fabricated forms. Most alloys are first hot-worked to break down the cast structure with the aim of achieving uniformity of both grain size as well as constituent size and distribution. Processing by cold-working may follow, particularly for sheet although it may be necessary to interrupt processing to give intermediate annealing treatments. Annealing usually involves heating the alloy to a temperature of 345 to 415° C and holding for times ranging from a few minutes up to 3 hours, depending upon alloy composition and section size. Cold-working is also important as the means of strengthening by work hardening those alloys which do not respond to age-hardening heat treatments.

In the hot-rolling of plate, the degree of working is uneven throughout

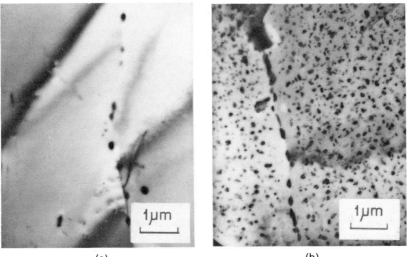

(a)   (b)

Fig. 3.5 Transmission electron micrographs showing formation of submicron particles of MnAl$_6$ in an Al-Zn-Mg alloy containing 0.3Mn. The ingots were homogenized at 500° C for 24 h following: a, a fast heating rate (500° C h^{-1}); b, a slow heating rate (50° C h^{-1}) (from Thomas, A. T., *Proceedings of 6th International Conference on Light Metals*, Leoben, Austria, Aluminium-Verlag, Düsseldorf, 1975)

the thickness, decreasing from surface to centre. Uniformity of work can be improved by increasing the reduction for each pass or by preforging or pressing the cast ingot before rolling. The problem of differential working throughout a section also applies with forgings and several techniques are used to control grain flow. In addition, it is necessary to exercise particular control of reheating cycles for alloys containing elements that assist in inhibiting recrystallization, so that coarse grains do not form in critically strained regions.

With extrusions, the avoidance of coarse, recrystallized grains is also a major objective as these tend to form around the periphery of sections which are very heavily worked as the material flows through the die (Fig. 3.6). The effect predominates at the back end of an extruded length because of the nature of flow during extrusion and is detrimental for several reasons, notably that the longitudinal tensile strength can be reduced by 10 to 15%. However, the heavy deformation associated with the process of extrusion is beneficial in producing a highly refined microstructure elsewhere in a section.

With sheet alloys used for some building materials and for high-strength aircraft panels, it is usual to roll-clad the surfaces with high-purity aluminium or an Al-1Zn alloy as a protection against atmospheric corrosion. This is arranged by attaching the cladding plates to each side of the freshly scalped ingot at the first rolling pass, taking care that the mating

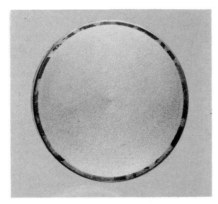

Fig. 3.6 Section through an extruded bar etched to show coarse recrystallized grains around the periphery (courtesy A. T. Thomas). ×1.5

surfaces are clean. Good bonding is achieved and each clad surface is normally 5% of the total thickness of the composite sheet (Fig. 3.7).

Fig. 3.7 Section of a high-strength alloy sheet roll-clad with pure aluminium (courtesy D. W. Glanvill). ×50

3.1.4 Thermal treatment

The function of thermal treatment is to develop a desired balance of mechanical properties required for consistent service performance. It will be clear from the above considerations that such consistency presupposes attainment of a satisfactory uniformity of microstructure in the preceeding stages of production of the wrought material. Also, annealing treatments, if required, are normally carried out within the temperature range 350 to 420° C. Attention must now be given to the processes of solution treatment, quenching and ageing, which are the three stages in strengthening alloys by precipitation hardening.

Solution treatment   Since the main purpose of this treatment is to obtain complete solution of alloying elements, it should ideally be carried out at a temperature within the single phase, equilibrium solid solution range for the alloy concerned. However, it is essential that alloys are not heated above the solidus temperature which will cause overheating, i.e. liquation of compounds and grain boundary regions with a subsequent adverse effect on ductility and other mechanical properties. Special problems are sometimes encountered with alloys based on the Al-Cu-Mg system in which adequate solution of the alloying elements is possible only if solution treatment is carried out within a few degrees of the solidus, and special care must be taken with the control of the furnace temperature.

Further precautions are necessary in solution treatment of hot-worked products, again to prevent the growth of coarse, recrystallized grains. Unnecessarily high temperatures and excessively long solution treatment times are to be avoided, particularly with extrusions or with forgings produced from extruded stock. Differential working that is common in such products is the reason why they are so sensitive to grain growth in localized regions.

The consequences of the introduction of hydrogen into molten aluminium through the surface reaction with water vapour were mentioned in Section 3.1.1. Such a reaction may also occur with solid aluminium during solution treatment leading to the adsorption of hydrogen atoms. These atoms can recombine at internal cavities to form pockets of molecular gas. Localized gas pressures can develop which, bearing in mind the relatively high plasticity of the metal at the solution treatment temperatures, may lead in turn to the irreversible formation of surface blisters (Fig. 3.8). Sources of the internal cavities at which blisters may form are unhealed porosity from the original ingot, intermetallic compounds that have cracked

Fig. 3.8 Blisters on the surface of an aluminium alloy component solution treated in a humid atmosphere (courtesy A. T. Thomas). × 3

during fabrication and, possibly, clusters of vacant lattice sites that may have formed when precipitates or compounds are dissolved. In these cases, the presence of blisters, while spoiling surface appearance, may have little effect upon mechanical properties of the components. However, blistering is often associated with overheating because the hydrogen can readily collect at liquated regions and this is a more serious problem requiring rejection of the affected material.

As it is difficult to eliminate internal cavities in wrought products, it is imperative that the water vapour content of furnace atmospheres be minimized. Where this is not possible, the introduction of a fluoride salt into the furnace during the heat treatment of critical components can be beneficial by reducing the surface reaction with water vapour.

Finally, reference should be made to the solution treatment of roll-clad sheet. Here it is necessary to avoid dissolved solute atoms, e.g. copper, diffusing from the alloy core through the high-purity aluminium or Al-Zn cladding thereby reducing its effect in providing protection against corrosion. Such a risk is particularly serious in thinner gauges of sheet. Strict control of temperature and time is necessary and such times should be kept to a minimum consistent with achieving full solution of the alloying elements in the core. The rate of heating to the solution treatment temperature is also important and it is customary to use a mixed nitrate salt bath for clad material because this rate is much faster than that found in air furnaces.

Quenching   After solution treatment, aluminium alloy components must be cooled or quenched, usually to room temperature, which is a straight-forward operation in principle since the aim is simply to achieve a maximum supersaturation of alloying elements in preparation for sub-sequent ageing. Cold water quenching is very effective for this and is frequently necessary in order to obtain adequate cooling rates in thicker sections. However, rapid quenching distorts thinner products such as sheet and introduces internal (residual) stresses into thicker products which are normally compressive at the surface and tensile in the core.

Residual stresses may cause dimensional instability, particularly when components have an irregular shape or when subsequent machining operations expose the underlying tensile stresses. What is also serious, is that the level of residual stresses may approach the yield stress in some high-strength alloys which, when superimposed upon normal assembly and service stresses, can cause premature failure. For products of regular section such as sheet, plate, and extrusions, the level of residual quenching stresses can be much reduced by stretching after quenching although this technique has limited use with sheet because it may cause an unacceptable reduction in thickness or gauge. In such cases roller levelling or flattening in a press may be a more satisfactory operation. The ageing treatment also allows some relaxation of stresses and reductions of between 20 and 40 % have been measured.

Quenching stresses will be reduced if slower rates of cooling are used and this alternative is particularly important in the case of forgings. Some alloys may be quenched with hot or boiling water, or even air-cooled after solution treatment, and still show an acceptable response to subsequent age hardening. The extent to which slower quenching rates can be tolerated is controlled by what is known as the quench sensitivity of the alloy concerned. During slow cooling there is a tendency for some of the solute elements to precipitate out as coarse particles which reduces the level of supersaturation and hence lowers the subsequent response of the alloys to age hardening. This effect is more pronounced in highly concentrated alloys and is aggravated by the presence of the submicron intermetallic compounds which provide interfaces for the heterogeneous nucleation of these precipitates (Fig. 3.9) during cooling. This behaviour is in fact the reverse of that occurring during the heating cycle when homogenizing ingots (Fig. 3.5). Changes may also occur in the region of grain boundaries which are a consequence of slow quenching. In particular, the segregation to the grain boundaries of solute elements such as copper may cause reduced toughness and susceptibility to intergranular corrosion in service.

The critical temperature range over which alloys display maximum quench sensitivity is 250 to 300° C and specialized techniques have been devised which allow rapid cooling through this range but still permit a reduction in residual stresses. One method which has been used for certain high-strength Al-Zn-Mg-Cu alloys is to quench into a fused salt bath at 180° C and hold for a time before cooling to room temperature. A second technique involves the use of proprietary organic liquids having an inverse solubility/temperature relationship. Whereas immersion of a hot body in boiling water generates a tenacious blanket of steam around the body thereby reducing the cooling rate in the critical temperature range, the

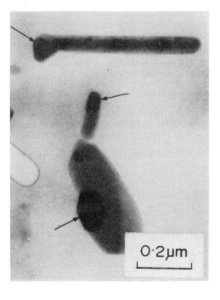

Fig. 3.9 Heterogeneous nucleation of the η phase ($MgZn_2$) on submicron intermetallic compounds (arrowed) during slow quenching of a commercial Al-Zn-Mg-Cu alloy (from Holl, H. A., *J. Inst. Metals,* **97**, 200, 1969)

0·2 μm

organic liquids are formulated to have the reverse effect. Initially the cooling rate is reduced by localized precipitation of a solute in the quenching medium after which it increases through the critical temperature range as the precipitate redissolves. The overall cooling rate is comparatively uniform and a desirable combination of stress-relief and high level of mechanical properties has been reported.

Ageing   Age hardening is the final stage in the development of properties in the heat treatable aluminium alloys. Metallurgical changes that occur during ageing have been discussed in Chapter 2. Some alloys will undergo ageing at room temperature (natural ageing) but most require heating for a time interval at one or more elevated temperatures (artificial ageing) which are usually in the range 100 to 190° C. Ageing temperatures and times are generally less critical than those in the solution treatment operation and depend upon the particular alloys concerned. Where single stage ageing is involved, a temperature is selected for which the ageing time to develop high-strength properties is of a convenient duration, e.g. 8 h corresponding to a working day or 16 h for an overnight treatment. Usually the only other stipulation is to ensure that the ageing time is sufficient to allow for the charge to reach the required temperature.

Multiple ageing treatments are sometimes given to certain alloys which have desirable effects on properties such as the stress-corrosion resistance. Such treatments may involve several days at room temperature followed by one or two periods at elevated temperatures. If alloys are slowly quenched after solution treatment, room temperature incubation may be critical because the lower supersaturation of vacancies alters precipitation kinetics. Sufficient time is required for the formation and growth of GP zones, particularly if these zones are to transform to another precipitate on subsequent elevated temperature ageing (Section 2.1.2). Similar considerations apply with respect to the rate of heating to the ageing temperature.

Thermomechanical processing   Some alloys show an enhanced response to hardening if they are cold-worked after quenching and prior to ageing and such treatments have been used for many years. A more recent development is called thermomechanical processing which involves plastic deformation at some stage during the actual precipitation reaction. Such treatments are carried out at a temperature high enough to develop a uniform distribution of dislocations and to stabilize this configuration by precipitates that are nucleated along the lines. This combination of precipitation and substructure hardening may enhance the strength and toughness of some alloys, and its effect on the fatigue properties of Al-Zn-Mg-Cu alloys was illustrated in Fig. 2.25.

3.2 Designation of alloys and tempers

3.2.1 Nomenclature of alloys

The selection of aluminium alloys for use in engineering has often been difficult because specifications and alloy designations have differed from country to country. Moreover, in some countries, the system used has been simply to number alloys in the historical sequence of their development rather than in a more logical arrangement. For these reasons, the introduction of an International Alloy Designation System (IADS) for wrought products in 1970 and its gradual acceptance by most countries is to be welcomed. The system is based on the classification used for many years by the Aluminum Association of the United States and it will be used when describing alloys in this book.

The IADS gives each wrought alloy a four digit number of which the first digit is assigned on the basis of the major alloying element(s) (Fig. 3.10). Hence there are the 1xxx series alloys which are unalloyed aluminium (with 99 % aluminium minimum), the 2xxx series with copper as the major alloying element, 3xxx series with manganese, 4xxx series with silicon, 5xxx series with magnesium, 6xxx with magnesium and silicon and the 7xxx series with zinc (and magnesium) as the major alloying elements. It should be noted however that wrought Al-Si alloys are used mainly for welding and brazing electrodes and brazing sheet.

The third and fourth digits are significant in the 1xxx series but not in other alloys. In the 1xxx series, the minimum purity of the aluminium is denoted by these digits, e.g. 1145 has a minimum purity of 99.45 %; 1200 has a minimum purity of 99.00 %. In all other series, the third and fourth digits have little meaning as they are nothing more than a serial number. Hence 3003, 3004, and 3005 are distinctly different Al-Mn alloys just as 5082 and 5083 denote two distinct Al-Mg alloys. The second digit indicates a close relationship, e.g. 5352 is closely related to 5052 and 5252, just as 7075 and 7475 differ only slightly in composition.

In Britain, it has been traditional to use three principal types of specifications to which aluminium and its alloys have been supplied:
(i) BS (British Standard) specifications for general engineering use
(ii) BS specifications for aeronautical use (designated as the L series)
(iii) DTD (Directorate of Technical Development) specifications issued by the Ministry of Technology for specialized aeronautical applications.
In addition there are several supplementary engineering specifications which cover other specialized alloys or those with limited use, while electrical applications are covered by a further six specifications.

The general engineering series is specified BS 1470–75, the six standards covering the different forms: plate, sheet, and strip (BS 1470); drawn tube (BS 1471); forging stock and forgings (BS 1472); rivet, bolt and screw stock (BS 1473); bars, extruded round tube and sections (BS 1474); wire (BS 1475). Every composition is denoted by a number which always indicates the same chemical composition irrespective of form or condition. Pure aluminium (99.99 % minimum content) is numbered 1 and the other grades by suffix 1A, 1B, and 1C. The alloys follow from 2 onwards with numerous omissions corresponding to obsolete alloys or numbers not now used. Non-heat treatable alloys are prefixed with the letter N and the heat treatable alloys with H. The various grades of aluminium (series 1) which

Aluminium alloy and temper designation systems

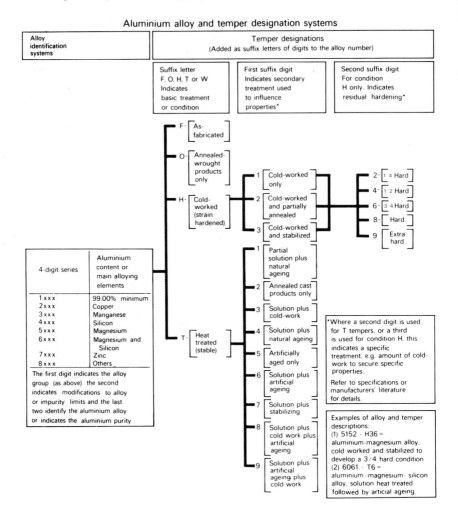

Fig. 3.10 Aluminium alloy and temper designation systems (courtesy Australasian Institute of Metals)

are not heat treatable do not, however, carry the N prefix. Other letters and figures indicate the form and condition of the material.

3.2.2 Temper or heat treatment nomenclatures

In order to specify the mechanical properties of an alloy and the way these properties were achieved, a system of temper nomenclature has also been adopted as part of the IADS. This takes the form of letters and digits that are added as suffixes to the alloy number. The system deals separately with the non-heat treatable, strain-hardening alloys on the one hand and heat treatable alloys on the other. The essential features of the system are outlined in Fig. 3.10 although recourse to detailed specifications or manufacturer's literature is recommended, particularly when several digits are included in the temper designation.

Alloys supplied in the as-fabricated or annealed conditions are designated with the suffixes F and O respectively. Strain hardening is a natural consequence of most working and forming operations on aluminium alloys. For the various grades of aluminium (1xxx series) and the non-heat treatable Al-Mn (3xxx) and Al-Mg (5xxx) alloys, this form of hardening augments the strengthening that arises from solid solution and dispersion hardening. Strain-hardened alloys are designated with the letter H. The first suffix digit indicates the secondary treatment: 1 is cold-worked only; 2 is cold-worked and partially annealed; 3 is cold-worked and stabilized. The second digit represents residual hardening, e.g. the severely cold-worked or fully hard condition is designated H18 which is equal to about a 75 % reduction in original cross-sectional area. The H16, H14 and H12 series of tempers are obtained with lesser degrees of cold-work and they represent what is commonly known as the three-quarter hard, half-hard and quarter-hard conditions, respectively. A combination of strain hardening and partial annealing is used to produce the H2 series of tempers. In these the products are cold-worked more than is required to achieve the desired mechanical properties and then reduced in strength by partial annealing. The H3 tempers apply only to Al-Mg alloys which have a tendency to soften with time at room temperature after strain hardening. This may be avoided by heating for a short time at an elevated temperature (120 to 175° C) to ensure completion of the softening process. This treatment provides stable mechanical properties and improves working characteristics.

A series of H temper designations having three digits have been assigned to wrought products as follows:

H111 applies to products which are strain-hardened less than the amount required for a controlled H11 temper.

H 112 applies to products which are strain-hardened less than the amount incidental to the shaping process. No special control is exerted over the amount of strain hardening or thermal treatment, but there are mechanical property limits and mechanical property testing is specified.

H121 applies to products which are strain-hardened less than the amount required for a controlled H12 temper.

H311 applies to products which are strain-hardened less than the amount required for a controlled H31 temper.

H321 applies to products which are strain-hardened less than the amount required for a controlled H32 temper.

In addition, temper designations H323 and H343 have been assigned to wrought products containing more than 4 % magnesium and apply to products that are specially fabricated to have acceptable resistance to stress-corrosion cracking.

A different system of nomenclature applies for heat treatable alloys. Tempers other than O are denoted by the letter T followed by one or more digits. The more common designations are: T4 which indicates that the alloy has been solution treated, quenched, and naturally aged; T5 in which the alloy has been rapidly cooled following processing at an elevated temperature, e.g. by extrusion, and then artificially aged; and T6 which denotes solution treatment, quenching and artificial ageing. For the T8 condition in which products are cold-worked between quenching and artificial ageing to improve strength, the amount of cold-work is indicated by a second digit, e.g. T85 means 5% cold-work.

Several designations involving additional digits have been assigned to stress-relieved tempers of wrought products:

Tx51 – stress-relieved by stretching. Applies to products that are stress-relieved after quenching by stretching the following amounts: plate 0.5 to 3% permanent set; rod, bar and shapes 1 to 3% permanent set. These products receive no further straightening after stretching. Additional digits are used in the designations for extruded rod, bar, shapes and tube as follows: Tx510 applies to products that receive no further straightening after stretching; Tx511 applies where minor straightening after stretching is necessary to comply with standard tolerances for straightness.

Tx52 – stress-relieved by compressing. Applies to products that are stress-relieved after quenching by compressing to produce a nominal permanent set of 2.5%.

Tx53 – stress-relieved by thermal treatment.

In the above designations the letter x represents digits 3, 4, 6, or 8 whichever is applicable.

In cases where wrought products may require heat treatment by the user, the following temper designations have been assigned:

T42 – solution treated, quenched and naturally aged

T62 – solution treated, quenched and artificially aged.

In the British system of nomenclature, the form of the product is denoted by the letters: B for bar and screw stock; C for clad plate, sheet or strip; E for bars, extruded round tube and sections; F for forgings and forging stock; G for wire; J for longitudinally welded tube; R for rivet stock; S for plate, sheet and strip; T for drawn tube. The condition of the product with regard to strain hardening or heat treatment is denoted by suffixes: M as manufactured; O annealed; H1 to H8 the degrees of strain hardening in increasing order of strength, with two additional categories, H68 applicable only to wire and H9 for extra-hard electrical wire. The heat treatment tempers are as follows:

TB – solution treated, quenched and naturally aged (formerly designated W).

TD – solution treated, cold-worked and naturally aged (applicable only to wires and formerly designated WD).

TE – artificially aged after cooling from a high temperature forming process (formerly P).

TF – solution treated, quenched and artificially aged (formerly WP).

TH – solution treated, quenched, cold-worked and artificially aged (applicable only to wire and formerly designated WDP).

The combination of letters and numbers enables an aluminium alloy to be identified and its form and condition described. For example HE9-TF is heat treatable composition 9 (Al-Mg-Si) in the form of bar, extruded round tube or section, in the fully heat-treated condition, i.e. solution treated, quenched and artificially aged. The equivalent designation in the IADS would be 6063-T6 which does not define the form of the product.

3.3 Work hardening of aluminium and its alloys

3.3.1 Strain-hardening characteristics

Strain hardening occurs during most working and forming operations and is the main method for strengthening aluminium and those alloys which do not respond to heat treatment. For heat treatable alloys, strain hardening may supplement the strength developed by precipitation hardening.

Tensile properties are the most affected and Fig. 3.11 shows work-hardening curves for 1100 aluminium and the alloys 3003 (Al-Mn) and 5052 (Al-Mg), the latter two being representative of the main classes of non-heat treatable alloys. Cold-working causes an initial rapid increase in yield strength, or proof stress, after which the increase is more gradual and roughly equals the change in tensile strength. These increases are obtained at the expense of ductility as measured by percentage elongation in a tensile test and also reduce formability in operations such as bending and stretch forming. For this reason, strain-hardened tempers are not usually employed when high levels of ductility and formability are required. However, it should be noted that certain alloys, e.g. 3003, exhibit better drawing properties in the cold-worked rather than the annealed condition and this is an important factor in making thin-walled beverage cans (Section 3.7.4).

The work-hardening characteristics of heat treatable alloys, in both the annealed and T4 tempers, are similar to those described above. As already mentioned, cold-working prior to ageing some of these alloys may cause additional strengthening (T8 temper). In the fully hardened, T6 temper, the increases in tensile properties by cold-working after ageing are comparatively small, except at very high strains, and are often limited by the poor workability of alloys in this condition. The principal use of this practice is for some extruded and drawn products such as wire, rod and tube which are cold-drawn after heat treatment to increase strength and improve surface finish. This applies particularly to products made from Al-Mg-Si alloys.

Work-hardening curves for annealed, recrystallized aluminium alloys, when plotted as a function of true stress and true strain, can be described by:

$$\sigma = k\varepsilon^n$$

where σ is true stress, k is the stress at unit strain, ε is the true or

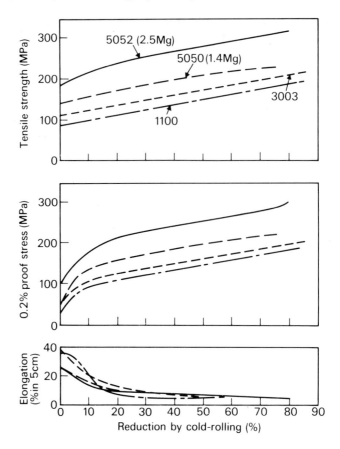

Fig. 3.11 Work-hardening curves for the alloys 1100 (99Al), 3003 (Al-1.2Mn), 5050 (Al-1.4Mg) and 5052 (Al-2.5Mg) (from Anderson, W. A., In: *Aluminum*, Volume 1, ed. K. Van Horn, American Society for Metals, Ohio, 1967)

logarithmic strain which is defined as $\ln A_o / A_f$ where A_o and A_f are the initial and final cross-sectional areas of a sample, and n is the work-hardening exponent. As the initial strengths of the alloys increase, k also increases whereas, for values of k in the stress range 175 to 450 MPa, n decreases from 0.25 to 0.17.

Rates of strain hardening can be calculated from the slopes of work-hardening curves. For non-heat treatable alloys initially in the cold-worked or hot-worked condition, these rates are substantially below those of annealed material. For the cold-worked tempers, this difference is caused by the strain necessary to produce the temper and, if this initial strain equals ε_0, then the equation for strain hardening becomes

$$\sigma = k(\varepsilon_0 + \varepsilon)^n.$$

A similar situation exists for products initially in the hot-worked condition. The strain hardening resulting from hot-working or forming is assumed to be equivalent to that achieved by a certain amount of cold-work. From a knowledge of the tensile properties of the hot-worked product, the amount of equivalent cold-work can be estimated by using the work-hardening curve for the annealed temper. By such procedures, it is usually possible to calculate work-hardening curves for hot-worked products that are in reasonable agreement with those for annealed products.

The work-hardening characteristics of aluminium alloys vary considerably with temperature. At cryogenic temperatures, strain hardening is greater than at room temperature, as shown in Fig. 3.12 which compares the work-hardening characteristics of the alloy 1100 at room temperature and $-196°C$. The gain in strength by working at $-196°C$ can be as much as 40% although there is a significant reduction in ductility. At elevated

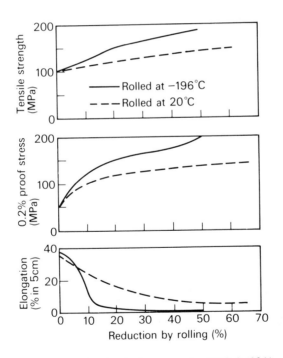

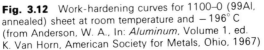

Fig. 3.12 Work-hardening curves for 1100–0 (99Al, annealed) sheet at room temperature and $-196°$ C (from Anderson, W. A., In: *Aluminum*, Volume 1, ed. K. Van Horn, American Society for Metals, Ohio, 1967)

temperatures, the work-hardening characteristics are influenced by both temperature and strain rate. Strain hardening decreases progressively as the working temperature is raised until a temperature is reached above which no effective hardening occurs due to dynamic recovery and recrystallization. This behaviour is important in commercial hot-working processes and it is necessary to determine the strength/temperature/time relationships when deforming different alloys in order to optimize these operations.

3.3.2 Substructure hardening

Aluminium has a high stacking fault energy and, during deformation, a cellular substructure is formed within the grains which causes some strengthening. In this regard, it has been shown that strength can be related to substructure by a Hall-Petch type of equation having the form

$$\sigma = \sigma_0 + k_1 d^{-m}$$

where σ is the yield strength, σ_0 is the frictional or Peierls stress, k_1 denotes the strength of the cell boundaries and m is an exponent that varies from 1 to 0.5. The substructure produced by working at relatively low temperatures is usually referred to as cells. These cells differ in orientation by only about $1°$ and have walls comprising tangled dislocations. On the other hand, deformation at higher temperatures produces 'subgrains' bounded by narrow, well-defined walls, and for which the misorientation is greater. The value of m changes from 1 to 0.5 if the alloys undergo the process of recovery which causes the substructure to change from cells to subgrains. The formation of subgrains is favoured if maximum substructure strengthening is desired.

3.3.3 Textures

Deformation of aluminium and its alloys proceeds by crystallographic slip that normally occurs on the $\{111\}$ planes in the $<110>$ directions. Large amounts of deformation at ambient temperatures lead to some strengthening through the development of textures, the nature of which depends in part on the mode of working. Aluminium wire, rod and bar usually have a 'fibre' texture in which the $<110>$ direction is parallel to the axis of the product, with a random orientation of crystal directions perpendicular to the axis. In rolled sheet, the texture that is developed may be described as a tube of preferred orientations linking $(110) < \bar{1}12 >$, $(112) < 11\bar{1} >$ and $(123) < \bar{2}\bar{1}1 >$.

When cold-worked aluminium or its alloys are recrystallized by annealing, new grains form with orientations that differ from those present in the cold-worked condition. Preferred orientation is much reduced but seldom eliminated and the annealing texture that remains has been extensively studied in rolled sheet. Some new grains form with a cube plane parallel to the surface and a cube edge aligned in the rolling direction, i.e.

(100) [001] texture, and other strain-free grains develop in which the rolling texture is retained. The texture and final grain size in recrystallized products are determined by the amount of cold-work, annealing conditions (i.e. rate of heating, annealing temperature and time), composition and the size and distribution of intermetallic compounds which tend to restrict grain growth.

Textures developed by cold-working cause directionality in certain mechanical properties. Texture hardening causes moderate increases in both yield and tensile strength in the direction of working and it has been estimated that, with an ideal fibre texture, the strength in the fibre direction may be 20 % higher than that for sheet with randomly oriented aggregates of grains. Forming characteristics of sheet may also be improved through an increase in the R-value for sheet which is the ratio of the strain in the width direction of a test piece to that in the thickness direction. A large R-value means there is a lack of deformation modes oriented to provide strain in the through-thickness direction. As a consequence, sheet will be more resistant to thinning during forming operations and this is desirable.

Preferred orientations in the plane of a sheet that are associated with textures may cause a problem known as earing (Fig. 3.13). Earing is wasteful of material because a larger blank than necessary must be used. Moreover it may lead to production problems due to difficulties in ejecting products after a pressing operation. Four ears usually form because of non-uniform plastic deformation along the rim of deep drawn products (Fig. 3.13). If the rolling texture is predominant, then ears will appear at 45° to the rolling direction of the sheet whereas, in the presence of the annealing (or cube) texture, they form in the direction of rolling and at right-angles to it. If there is a desirable balance between the two textures, there will be either eight small ears or none at all. Earing may be minimized by careful control of rolling and annealing schedules, e.g. sheet is sometimes cross-rolled so that textures are less clearly defined.

Crystallographic textures should not be confused with mechanical fibring that occurs because of changes in grain shape, banding of small

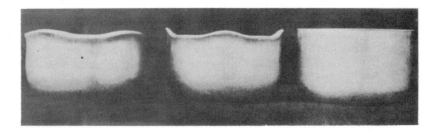

Fig. 3.13 Deep drawn aluminium cups showing 45° earing, 0–90° earing and no earing with respect to the rolling direction (from Anderson, W. A., In: *Aluminum*, Volume 1, ed. K. Van Horn, American Society for Metals, Ohio, 1967)

grains or the alignment of particles in worked alloys (Figs. 2.17 and 2.18). As mentioned in Sections 2.4.1 and 2.4.2, these effects also contribute to the anisotropy of mechanical properties and they are often more important than crystallographic texture.

3.3.4 Secondary work-hardening effects

Aluminium products formed by stretching, bending or drawing sometimes develop a roughened surface and this effect is known as 'orange peeling'. It is common to other materials and is caused by the presence of coarse grains at the surface.

A problem that occurs in a few aluminium alloys, but not in other non-ferrous materials, is the formation of stretcher strain markings, or Lüders lines, during the forming or stretching of sheet. These markings occur in one of two forms and may give rise to differences in surface topography of sheet during drawing and stretch-forming operations. One type occurs in annealed or heat treated solid solution alloys, notably Al-Mg, and is produced when yielding takes place in some parts of a sheet but not in others. It is similar in origin to the well-known Lüders lines that may form during the deformation of certain sheet steels. The second type is associated with the Portevin-LeChatelier effect and it produces uneven or serrated yielding during a tensile test. Diagonal bands appear, oriented approximately $50°$ to the tension axis, which move up or down during stretching and terminate at the grips. This type of marking is rarely observed in commercial forming operations but may appear when strain-hardened sheet or plate is stretched to produce flatness.

Stretcher strain markings are generally undesirable because they cause uneven and roughened surfaces. They can be avoided by forming with strain-hardened rather than annealed sheet, providing the material has adequate ductility. The formation of these markings can also be avoided or minimized by forming or working at temperatures above $150°$ C.

3.4 Non-heat treatable alloys

Wrought compositions that do not respond to strengthening by heat treatment mainly comprise the various grades of aluminium as well as alloys with manganese or magnesium as the major additions, either singly or in combination (Table 3.1). Strength is developed by strain hardening, usually by cold-working during fabrication, in association with dispersion hardening (Al-Mn) or solid solution hardening (Al-Mg) or both (Al-Mn-Mg). The miscellaneous alloys in the 8xxx series mostly do not respond to heat treatment and are used for specific applications such as bearings and bottle caps. An Al-1Zn alloy designated 7072 serves as cladding to protect a number of other alloys, e.g. 2219, 7075, from corrosion (Section 3.1.3) and for some fin stock. Several Al-Si alloys in the 4xxx series are available but they are used mainly for welding electrodes, e.g. 4043, or as brazing rods, e.g. 4343. Typical mechanical properties are given in Table 3.2 although it

Table 3.1 Compositions of selected non-heat treatable wrought aluminium alloys

IADS designation	Si	Fe	Cu	Mn	Mg	Zn	Cr	Ti	Other
1100	1.0 Si +	Fe	0.05–0.20	0.05		0.10		0.05	Al min 99.0
1200	1.0 Si +	Fe	0.05	0.05		0.10		0.03	Al min 99.0
1145	0.55 Si +	Fe	0.05	0.05	0.05	0.05			Al min 99.45
1199	0.006	0.006	0.006	0.002	0.006	0.006		0.002	Al min 99.99
3003	0.6	0.7	0.05–0.20	1.0–1.5		0.10			
3004	0.30	0.7	0.25	1.0–1.5	0.8–1.3	0.25			
3005	0.6	0.7	0.30	1.0–1.5	0.20–0.6	0.25	0.10		
5005	0.30	0.7	0.20	0.20	0.50–1.1	0.25	0.10		
5050	0.40	0.7	0.20	0.10	1.1–1.8	0.25	0.10		
5052	0.25	0.40	0.10	0.10	2.2–2.8	0.10	0.15–0.35		
5454	0.25	0.40	0.10	0.50–1.0	2.4–3.0	0.25	0.05–0.20	0.20	
5456	0.25	0.40	0.10	0.50–1.0	4.7–5.5	0.25	0.05–0.20	0.20	
5082	0.20	0.35	0.15	0.15	4.0–5.0	0.25	0.15	0.10	
5083	0.40	0.40	0.10	0.40–1.0	4.0–4.9	0.25	0.05–0.25	0.15	
5086	0.40	0.50	0.10	0.20–0.7	3.5–4.5	0.25	0.05–0.25	0.15	
8001	0.17	0.45–0.7	0.15	0.10		0.10	0.05		0.09–1.3 Ni
8011	0.50–0.9	0.6–1.0	0.10	0.10	0.05	0.05	0.05	0.08	5.5–7.0 Sn
8280	1.0–2.0	0.7	0.7–1.3						0.20–0.7 Ni
8081	0.7	0.7	0.7–1.3	0.10		0.05		0.10	18–22 Sn

Compositions are in % maximum by weight unless shown as a range or a minimum

should be noted that the alloys are mainly used for applications in which other properties are the prime consideration. The commercial forms in which they are available and some typical applications are also included in Table 3.2. All are readily weldable and have a high resistance to corrosion in most media.

Some of the essential features of the individual classes of alloys are discussed below. More details concerning special products and applications, e.g. electrical conductors, are considered in Section 3.7.

3.4.1 Super-purity and commercial-purity aluminium (1xxx series)

This group includes super-purity (SP) aluminium (99.99 %) and the various grades of commercial-purity (CP) aluminium containing up to 1 % of impurities or minor additions. The materials have been utilized as wrought products since the industry was first developed and the CP grades are available in most forms. Tensile properties are low and annealed SP aluminium has a proof stress of only 10 to 11 MPa. Applications include electrical conductors for which there are several compositions (see Section 3.7.1), chemical process equipment, foil (Section 3.7.4) and architectural products requiring decorative finishes.

3.4.2 Al-Mn and Al-Mn-Mg alloys (3xxx series)

Commercial Al-Mn alloys contain up to 1.25 % manganese although the maximum solid solubility of this element in aluminium is as high as 1.82 %. This limitation is imposed because the presence of iron as an impurity reduces the solubility and there is a danger that large, primary particles of $MnAl_6$ will form with a disastrous effect on local ductility. The only widely used binary Al-Mn alloy is 3003 which is supplied as sheet. The presence of finer manganese-containing intermetallic compounds confers some dispersion hardening, and the tensile strength of annealed 3003 is typically 110 MPa compared with 90 MPa for CP aluminium (1100), with corresponding increases in the work-hardened tempers. The addition of magnesium, as in the more commonly used alloy 3004, gives a further increase in tensile strength to 180 MPa for the annealed condition. Magnesium also raises the recrystallization temperature by some 50 to 60° C. In general, the 3xxx series of alloys is used when moderate strength combined with high ductility and excellent corrosion resistance is required. They find particular use for the manufacture of cans for the beverage industry (Section 3.7.4) as well as for cooking utensils and roofing sheet.

3.4.3 Al-Mg alloys (5xxx series)

Aluminium and magnesium form solid solutions over a wide range of compositions and wrought alloys containing from 0.8 % to slightly more

Table 3.2 Typical mechanical properties and applications of selected non-heat treatable wrought aluminium alloys

IADS designation	Temper	0.2% proof stress (MPa)	Tensile strength (MPa)	Elongation (% in 50 mm)	Typical applications
1100	O	35	90	35	Sheet, plate, tube, wire, spun hollow-ware, food equipment
	H18	150	165	5	
1145	O	35	75	40	Foil, sheet, semi-rigid containers
	H18	115	145	5	
1199	O	10	45	50	Electrical and electronic foil
	H18	110	115	5	
3003	O	25	110	30	Sheet, plate, foil, rigid containers, cooking utensils, tubes
	H18	185	200	4	
3004	O	70	180	20	Sheet, plate, rigid containers
	H38	250	280	5	
3005	O	55	130	25	Higher strength foil, roofing sheet
	H18	225	240	4	
5005	O	40	125	30	General sheet material, high-strength foil, electrical conductor wire
	H18	195	200	4	
	H38	185	200	5	
5050	O	55	145	24	Sheet and tube, rip tops for cans
	H38	200	220	6	
5052	O	90	195	25	Sheet, plate, tubes, marine fittings
	H38	255	270	7	
5454	O	120	250	22	Special purpose sheet, plate, extrusions, pressure vessels, marine applications such as hulls and superstructures, dump truck bodies, cryogenic structures
	H34	240	305	10	
5456	O	160	310	24	
	H24	280	370	12	
5083	O	145	290	22	
	H343	280	360	8	
5086	O	115	260	22	
	H34	255	325	10	
8001	O	40	110	30	Sheet, tubing for water-cooled nuclear reactors
	H18	185	200	4	

than 5% magnesium are widely used. Strength values in the annealed condition range from 40 MPa yield and 125 MPa tensile for Al-0.8Mg (5005) to 160 MPa yield and 310 MPa tensile for the strongest alloy 5456 (Table 3.2). Elongations are relatively high and usually exceed 25%. The alloys work harden rapidly at rates that increase as the magnesium content is raised (Fig. 3.11). Fully work-hardened 5456 has yield strength of 300 MPa and a tensile strength of 385 MPa with an elongation of 5%.

The alloys may exhibit some instability in properties which is manifest in two ways. If the magnesium content exceeds 3 to 4%, there is a tendency for the β-phase, Mg_5Al_8, to precipitate in slip bands and grain boundaries which may lead to intergranular attack and stress-corrosion cracking in corrosive conditions. Precipitation of β occurs only slowly at ambient temperatures but is accelerated if the alloys are in a heavily worked condition, or if the temperature is raised. Small additions of chromium and manganese, which are present in most alloys and raise the recrystallization temperatures, also increase tensile properties for a given magnesium content. This offers the prospect of using alloys with reduced magnesium contents if precipitation of β is to be avoided. For example, alloy 5454 contains 2.7%Mg, 0.7%Mn and 0.12%Cr and has tensile properties similar to those expected from a binary alloy having as much as 4% magnesium.

The second problem is that work-hardened alloys may undergo what is known as age softening at ambient temperatures. Over a period of time, the tensile properties fall due to localized recovery within the deformed grains and, as mentioned in Section 3.2.2, a series of H3 tempers has been devised to overcome this effect. These tempers involve cold-working to a level slightly greater than desired and then stabilizing by heating to a temperature of 120 to 150° C. This lowers the tensile properties to the required level and stabilizes them with respect to time. The treatment also reduces the tendency for precipitation of β in the higher magnesium alloys.

Al-Mg alloys are widely used for welded applications. In transportation structural plate is used for dump truck bodies, large tanks for carrying petrol, milk and grain, and pressure vessels, particularly where cryogenic storage is involved. Their high corrosion resistance makes them suitable for the hulls of small boats and for the superstructures of ocean-going vessels. In addition, they polish to a bright surface finish, particularly if made from high-purity aluminium, and are used for automotive trim and architectural components.

3.4.4 Miscellaneous alloys (8xxx series)

This series contains several dilute alloys, e.g. 8001 (A1-1.1Ni-0.6Fe) which is used in nuclear energy installations where resistance to corrosive attack by water at high temperatures and pressures is the desired characteristic. Its mechanical properties resemble 3003. Alloy 8011 (Al-0.75Fe-0.7Si) is used for bottle caps because of its good deep drawing qualities and several other dilute compositions are included in the range of electrical conductor materials (Section 3.7).

Alloys such as 8280 and 8081 serve an important role as bearing alloys based on the Al-Sn system that are now widely used in motor cars and trucks, particularly where diesel engines are involved. These alloys are considered in some detail in Section 3.7.2.

3.5 Heat treatable alloys

Wrought alloys that respond to strengthening by heat treatment are covered by the three series 2xxx (Al-Cu, Al-Cu-Mg), 6xxx (Al-Mg-Si) and 7xxx (Al-Zn-Mg, Al-Zn-Mg-Cu). All depend on age hardening to develop enhanced strength properties and they can be classified into two groups: those that have medium strength and are readily weldable (Al-Mg-Si and Al-Zn-Mg), and the high-strength alloys that have been developed primarily for aircraft construction (Al-Cu, Al-Cu-Mg, and Al-Zn-Mg-Cu) most of which have very limited weldability. Compositions of representative commercial alloys are shown in Table 3.3 and Table 3.4 gives typical properties, the forms in which they are available, together with some common applications.

3.5.1 Al-Cu alloys (2xxx series)

Although the complex changes that occur during the ageing of Al-Cu alloys have been studied in greater detail than any other system, there are actually few commercial alloys based on the binary system. Alloy 2011 (Al-5.5Cu) is used where good machining characteristics are required, for which it contains small amounts of the insoluble elements lead and bismuth that form discrete particles in the microstructure and assist with chip formation. Alloy 2025 is used for some forgings although it has largely been superseded by 2219 (Al-6.3Cu) which has a more useful combination of properties and is also available as sheet, plate and extrusions. Alloy 2219 has relatively high tensile properties at room temperature together with good creep strength at elevated temperatures and high toughness at cryogenic temperatures. In addition, it can be welded and has been used for fuel tanks for storing liquified gases that serve as propellants for missiles and space vehicles. Response to age hardening is enhanced by strain hardening prior to artificial ageing (T8 temper) and the yield strength may be increased by as much as 35 % as compared with the T6 temper (Table 3.4).

A modified version of 2219 has been developed in the United States in order to meet a requirement for an increase of 10 to 15 % in tensile properties. This alloy is designated 2021 and, as rolled plate, has a yield strength of 435 MPa, tensile strength 505 MPa and elongation of 9 %, with no reported sacrifice of weldability or toughness at cryogenic temperatures. Increased strength has been achieved by minor additions of 0.15 % cadmium and 0.05 % tin which have the well-known effect of refining the

Table 3.3 Compositions of selected heat treatable wrought aluminium alloys

IADS designation	Si	Fe	Cu	Mn	Mg	Zn	Cr	Ti	Other
2011	0.40	0.7	5.0–6.0			0.30	0.10	0.15	0.2–0.6Bi, 0.2–0.6Pb
2014	0.50–1.2	0.7	3.9–5.0	0.40–1.2	0.20–0.8	0.25	0.10	0.15	0.2Zr+Ti
2017	0.2–0.8	0.7	3.5–4.5	0.4–1.0	0.4–0.8	0.25			0.2Zr+Ti
2618	0.10–0.25	0.9–1.3	1.9–2.7		1.3–1.8	0.10		0.04–0.10	0.9–1.2Ni
2219	0.20	0.30	5.8–6.8	0.20–0.40	0.02	0.10		0.02–0.10	0.05–0.15V, 0.10–0.25Zr
2021	0.20	0.30	5.8–6.8	0.20–0.40	0.02	0.10		0.02–0.10	0.10–0.25Zr, 0.05–0.20Cd
2024	0.50	0.50	3.8–4.9	0.30–0.9	1.2–1.8	0.25	0.10	0.15	
2124	0.20	0.30	3.8–4.9	0.30–0.9	1.2–1.8	0.25	0.10	0.15	
2025	0.50–1.2	1.0	3.9–5.0	0.40–1.2	0.05	0.25	0.10	0.15	0.20Zr+Ti
2036	0.50	0.50	2.2–3.0	0.10–0.40	0.30–0.6	0.25	0.10	0.15	0.20Zr+Ti
2048	0.15	0.20	2.8–3.8	0.20–0.6	1.2–1.8	0.25		0.10	
2020	0.40	0.40	4.0–5.0	0.30–0.8	0.03	0.25		0.10	0.9–1.7Li, 0.10–0.25Cd
6063	0.20–0.6	0.35	0.10	0.10	0.45–0.9	0.10	0.10	0.10	
6463	0.20–0.6	0.15	0.20	0.05	0.45–0.9	0.05			
6061	0.40–0.8	0.7	0.15–0.40	0.15	0.8–1.2	0.25	0.04–0.35	0.15	
6151	0.6–1.2	1.0	0.35	0.20	0.45–0.8	0.25	0.15–0.35	0.15	
6351	0.7–1.3	0.50	0.10	0.40–0.8	0.40–0.8	0.20		0.20	
6262	0.40–0.8	0.7	0.15–0.40	0.15	0.8–1.2	0.25	0.04–0.14	0.15	0.40–0.7Bi, 0.40–0.7Pb
6009	0.6–1.6	0.50	0.15–0.6	0.2–0.8	0.40–0.8	0.25	0.1	0.10	
6010	0.8–1.2	0.50	0.15–0.6	0.2–0.8	0.6–1.0	0.25	0.1	0.10	

IADS designation	Si	Fe	Cu	Mn	Mg	Zn	Cr	Ti	Other
7001	0.35	0.040	1.6–2.6	0.20	2.6–3.4	6.8–8.0	0.18–0.35	0.20	0.10–0.20Zr
7004	0.25	0.35	0.05	0.20–0.7	1.0–2.0	3.8–4.6	0.05	0.05	0.08–0.20Zr
7005	0.35	0.40	0.10	0.20–0.7	1.0–1.8	4.0–5.0	0.06–0.20	0.01–0.06	
7009	0.20	0.20	0.6–1.3	0.10	2.1–2.9	5.5–6.5	0.10–0.25	0.20	0.25–0.40Ag
7010	0.10	0.15	1.5–2.0	0.30	2.2–2.7	5.7–6.7	0.05		0.11–0.17Zr
7039	0.30	0.40	0.10	0.10–0.40	2.3–3.3	3.5–4.5	0.15–0.25	0.10	
7049	0.25	0.35	1.2–1.9	0.20	2.0–2.9	7.2–8.2	0.10–0.22	0.10	
7050	0.12	0.15	2.0–2.6	0.10	1.9–2.6	5.7–6.7	0.04	0.06	0.08–0.15Zr
7075	0.40	0.50	1.2–2.0	0.30	2.1–2.9	5.1–6.1	0.18–0.28	0.20	0.25Zr +Ti
7475	0.10	0.12	1.2–1.9	0.06	1.9–2.6	5.2–6.2	0.18–0.25	0.06	
7178	0.40	0.50	1.6–2.4	0.30	2.4–3.1	6.3–7.3	0.18–0.35	0.20	
7079	0.30	0.40	0.40–0.8	0.10–0.30	2.9–3.7	3.8–4.8	0.10–0.25	0.10	

Compositions are in % maximum by weight unless shown as a range

Table 3.4 Typical mechanical properties and applications of selected heat treatable wrought aluminium alloys

IADS designation	Temper	0.2% proof stress (MPa)	Tensile strength (MPa)	Elongation (% in 50 mm)	Typical applications
2011	T6	295	390	17	Screw machine parts
2014	T6	410	480	13	Aircraft structures
2017	T4	275	425	22	Screw machine fittings
2618	T61	330	435	10	Aircraft parts and structures for use at elevated temperatures. 2219 is weldable
2219	T62	290	415	10	
2219	T87	395	475	10	
2024	T4	325	470	20	Aircraft structures and sheet. Truck wheels
	T6	395	475	10	
	T8	450	480	6	
2124	T8	440	490	8	Aircraft structures
2025	T6	255	400	19	Forgings, aircraft propellors
2036	T4	195	340	24	Automotive body panels
2048	T85	440	480	10	Aircraft structures
2020	T6	530	580	7	Aircraft structures
6063	T6	215	240	12	Architectural extrusions, pipes
6061	T6	275	310	12	Welded structures
6151	T6	295	330	17	Medium-strength forgings
7001	T6	625	675	9	High-strength aircraft structures
7004	T6	340	400	12	Medium-strength welded structures
7005	T53	345	395	15	
7009	T6	470	535	12	Aircraft structures
7010	T6	485	545	12	
7039	T61	345	415	13	Medium-strength welded structures
7049	T73	470	530	11	
7050	T736	510	550	11	High-strength aircraft structures including extrusions, forgings and sheet
7075	T6	500	570	11	
	T73	430	500	13	
	T76	470	540	12	
7475	7651	560	590	12	
7178	T6	540	610	10	
7079	T6	470	540	14	Aircraft forgings (now superseded)

size of the θ' transition precipitate which forms on ageing in the medium temperature range ($\sim$ 130 to 200°C). However, the toxicity of cadmium requires that casting operations be carried out under carefully controlled conditions.

The role of minor or trace element effects in modifying the nucleation and growth of precipitates during ageing has also been exploited in some experimental British alloys which are known by the acronym Almagem. One such alloy has the nominal composition Al-6Cu-0.3Mg-0.6Mn-0.2Ge-0.1Si with a low iron content. It is believed that the magnesium and germanium interact strongly with vacancies and stimulate nucleation of the phase θ'. Sufficient manganese is added to remove the iron into the compound $(Mn,Fe)Al_6$ which allows the small amount of silicon to segregate to the θ'/matrix interface and reduce the rate of coarsening of this phase at elevated temperatures. Creep resistance is improved and typical mechanical properties for the alloy aged at 170 to 190°C are: 0.2% proof stress 425 MPa, tensile strength 500 MPa, elongation 12%, with a total plastic strain as low as 0.04% after creep testing for 1000 h at 150°C under a stress of 185 MPa.

3.5.2 Al-Cu-Mg alloys (2xxx series)

These alloys date from the accidental discovery of the phenomenon of age hardening by Alfred Wilm working in Berlin in 1906 who was seeking to develop a stronger aluminium alloy to replace brass for the manufacture of cartridge cases. His work led to the production of an alloy known as Duralumin (Al-3.5Cu-0.5Mg-0.5Mn) which was quickly used for structural members for Zeppelin airships, and later for aircraft. A modified version of this alloy (2017) is still used, mainly for rivets, and several other important alloys have been developed which are now widely used for aircraft construction. An example is 2014 (Al-4.4Cu-0.5Mg-0.9Si-0.8Mn) in which higher strengths have been achieved because the relatively high silicon content increases the response to hardening on artificial ageing. Typical tensile properties are: 0.2% proof stress 320 MPa and tensile strength 485 MPa. Another alloy 2024, in which the magnesium content is raised to 1.5% and the silicon content is reduced to impurity levels, undergoes significant hardening by natural ageing at room temperature and is frequently used in T3 or T4 tempers. It also has a high response to artificial ageing particularly if cold-worked prior to ageing at around 175°C, e.g. 0.2% proof stress 490 MPa and tensile strength 520 MPa for the T86 temper.

These and other 2xxx alloys in the form of sheet are normally roll-clad with aluminium or Al-1Zn in order to provide protection against corrosion, and the tensile properties of the composite product may be some 5% below those for the unclad alloy. Much greater reductions in strength may occur under fatigue conditions. For example, cladding 2014 sheet may reduce the fatigue strength by as much as 50% if tests are conducted in air under reversed plane bending. Differences between clad and unclad alloys are much less under axial loading conditions, or if the materials are tested as part of a structural assembly. Under corrosion-fatigue conditions, the strength of unclad sheet may fall well below that of the clad alloy. The adverse effect of cladding on fatigue strength in air is due mainly to the ease by which cracks can be initiated in the soft surface layers and a number of harder cladding materials are being investigated.

Microstructural features that influence toughness and ductility have been considered in Section 2.4.2. Fig. 3.14 shows that, for equal values of yield strength, alloys in the 2xxx series have lower fracture toughness than those of the 7xxx series (Al-Zn-Mg-Cu). This is attributed to the larger sizes of intermetallic compounds in the 2xxx alloys. Improvements in both fracture toughness and ductility can be obtained by reducing the levels of iron and silicon impurities as well as that of copper, all of which favour

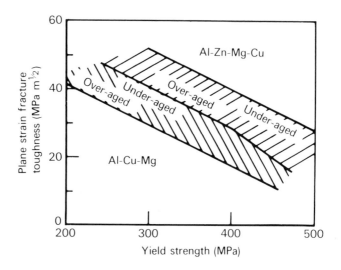

Fig. 3.14   Relationships of plane strain fracture toughness to yield strength for the 2xxx and 7xxx series of alloys (from Develay, R., *Metals and Materials*, **6**, 404, 1972)

formation of large, brittle compounds in the cast materials. This has led to the development of alloys such as 2124 (iron 0.3 % maximum, silicon 0.2 % maximum, as compared with levels up to 0.5 % for each of these elements in 2024) and 2048 (copper reduced to 3.3 % and iron and silicon 0.20 % and 0.15 % maximum, respectively). Some comparative properties are shown in Fig. 2.20 and Table 3.5. High-toughness versions of older alloys such as 2024 are now being used as sheet, plate and forgings in several modern aircraft.

Relationships between rate of growth of fatigue cracks and stress intensity for the alloys 2024-T3 and 7075-T6 are shown in Fig. 3.15. Other 2xxx series alloys show similar rates of crack propagation to 2024-T3 over

Table 3.5 Effect of purity on the fracture toughness of some high-strength wrought aluminium alloys (from Speidel, M.O., *Met. Trans.*, **6A**, 631, 1975)

Alloy and temper	% Fe maximum	% Si maximum	0.2% proof stress (MPa)	Tensile strength (MPa)	Fracture toughness (MPa m$^{1/2}$) Longitudinal	Short transverse
2024-T8	0.50	0.50	450	480	22–27	18–22
2124-T8	0.30	0.20	440	490	31	25
2048-T8	0.20	0.15	420	460	37	28
7075-T6	0.50	0.40	500	570	26–29	17–22
7075-T73	0.50	0.40	430	500	31–33	20–23
7175-T736	0.20	0.15	470	540	33–38	21–29
7050-T736	0.15	0.12	510	550	33–39	21–29

most of the range of test conditions. In general, these alloys have rates of crack growth that are close to one-third those observed in the 7xxx series alloys.

It is now common to use pre-cracked specimens to assess comparative resistance of alloys to stress-corrosion cracking since this type of test avoids uncertainties associated with crack initiation. Relationships between crack

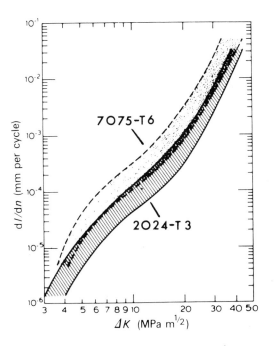

Fig. 3.15 Comparative fatigue crack growth rates for 2024-T3 and 7075-T6 in air of varying humidity (from Hahn, G. T. and Simon, R., *Eng. Fract. Mech.*, **5**, 523, 1973)

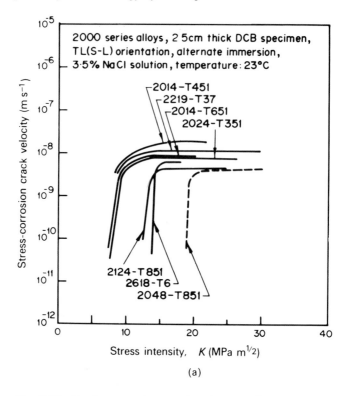

Fig. 3.16 Crack propagation rates in stress-corrosion tests using pre-cracked specimens of high-strength aluminium alloys exposed to an aqueous solution of 3.5% NaCl: a, 2xxx series alloys – 25 mm thick, double cantilever beam (DCB) specimens, short-transverse direction of plate, wet with solution twice a day

growth rate and stress intensity in tests on several 2xxx series alloys are shown in Fig. 3.16a. It is clear that there is a large variation in threshold stress intensities for different alloys, while the crack velocity plateaus are similar. Thus the ranking of these alloys is best obtained from measurements of the threshold stress intensities (K_{Iscc} values). In the naturally aged tempers T3 and T4, the 2xxx series are prone to stress-corrosion cracking. The alloy 2014 in the T6 temper is also susceptible whereas the more recent alloy 2048 in the T8 temper, is much more resistant to this form of cracking.

Interest in supersonic aircraft such as Concorde stimulated a need for a sheet alloy with improved creep strength on prolonged exposure, e.g. 50 000 h, at temperatures of around 120°C. Such a material was developed from a forging alloy known widely as RR58 (2618) which itself had been adapted from an early casting alloy (Al-4Cu-1.5Mg-2Ni) known originally as Y alloy. The alloy 2618 has the nominal composition Al-2.2Cu-1.5Mg-

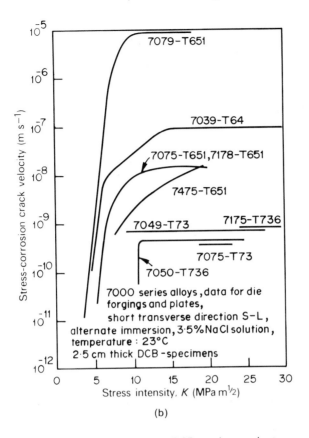

Fig. 3.16b, 7xxx series alloys–DCB specimens, short-transverse direction of die forgings and plate, alternate immersion tests (from Speidel, M, O., *Met. Trans.*, **6A**, 631, 1975)

1Ni-1Fe, in which the copper and magnesium contribute to strengthening through age hardening, whereas nickel and iron form the intermetallic compound $FeNiAl_9$ which causes dispersion hardening and assists in stabilizing the microstructure. A recent refinement has been the addition of 0.2 % silicon which both increases the hardening associated with the first stage in the ageing process (GPB zones) and promotes formation of a more uniform dispersion of the S phase (Table 2.2). Both these changes improve the creep properties of the alloy on long-term exposure at temperatures of 120 to 150° C.

Lower strength Al-Cu-Mg alloys are being investigated as possible sheet materials for automotive applications and these are discussed in Section 3.7.3. One example is 2036 which has nominal copper and magnesium contents of 2.5 % and 0.45 % respectively.

3.5.3 Al-Mg-Si alloys (6xxx series)

Al-Mg-Si alloys are widely used as medium-strength structural alloys which have the additional advantages of good weldability, corrosion resistance, and immunity to stress-corrosion cracking. Just as the 5xxx series of alloys comprise the bulk of sheet products, the 6xxx series are used for the majority of extrusions, with smaller quantities being available as sheet and plate (Fig. 3.17). Magnesium and silicon are added either in balanced amounts to form quasi-binary Al-Mg_2Si alloys (Mg:Si 1.73:1), or with an excess of silicon above that needed to form Mg_2Si. The commercial alloys may be divided into three groups.

The first group comprises alloys with balanced amounts of magnesium and silicon adding up to between 0.8 % and 1.2 %. These alloys can be readily extruded and offer a further advantage in that they may be quenched at the extrusion press when the product emerges hot from the die, thereby eliminating the need to solution treat as a separate operation. Quenching is normally achieved by means of water sprays, or by leading the product through a trough of water. Thin sections (< 3 mm) can be air-cooled. Moderate strength is developed by age hardening at 160 to 190° C and one alloy, 6063, is perhaps the most widely used of all Al-Mg-Si alloys. In the T6 temper, typical tensile properties are 0.2 % proof stress 215 MPa and tensile strength 245 MPa. These alloys find particular application for architectural and decorative finishes and, in this regard, they respond well to clear or colour anodizing as well as to the application of other surface finishes. A high-purity version, 6463, in which the iron content is kept to a

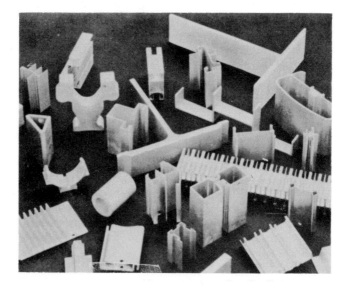

Fig. 3.17   Examples of extruded sections made from Al-Mg-Si alloys

low level (< 0.15%), responds well to chemical brightening and anodizing for use as automotive trim.

The other two groups contain magnesium and silicon in excess of 1.4%. They develop higher strength on ageing and, because they are more quench sensitive, it is usually necessary to solution treat and water quench as separate operations after extrusion. One group, which is particularly popular in North America, has balanced compositions and a common example is 6061 (Al-1Mg-0.6Si) to which is added 0.25% copper to improve mechanical properties together with 0.2% chromium to offset any adverse effect copper may have on corrosion resistance. These alloys are widely used as general purpose structural materials. The alloys in the other group contain silicon in excess of that needed to form Mg_2Si and the presence of this excess silicon promotes an additional response to age hardening by both refining the size of the Mg_2Si particles and precipitating as silicon. However, it may also reduce ductility and cause intergranular embrittlement which is attributed in part to the tendency for silicon to segregate to the grain boundaries in these alloys. The presence of chromium (6151) and manganese (6351) help to counter this effect by promoting fine grain size and inhibiting recrystallization during solution treatment. The alloys are used as extrusions and forgings.

Al-Mg-Si alloys are normally aged at about 170° C. During commercial processing, there may be a delay at room temperature between quenching and artificial ageing which may modify the mechanical properties that are developed. In alloys containing more than 1% Mg_2Si, a delay of 24 h causes a reduction of up to 10% in tensile properties as compared with the properties obtained by ageing immediately. However, such a delay can enhance the tensile properties developed in compositions containing less than 0.9% Mg_2Si. These effects have been attributed to clustering of solute atoms and vacancies that occurs at room temperature and to the fact that the GP zones solvus (Section 2.1.2) is above 170°C for the more highly alloyed compositions. With these alloys, the precipitate which develops directly from the clusters formed at room temperature is coarser than that developed in alloys aged immediately after quenching, with a consequent adverse effect on tensile properties. The reverse occurs in alloys containing less than 0.9% Mg_2Si. The addition of 0.25% copper lessens effects of delays at room temperature because copper reduces the rate of natural ageing in Al-Mg-Si alloys.

As with the 2xxx series, there is an Al-Mg-Si alloy (6262) containing additions of lead and bismuth to improve machining characteristics. Although the machinability of this alloy is below the Al-Cu alloy 2011, it is not susceptible to stress-corrosion cracking and is preferred for more highly stressed fittings.

3.5.4 Al-Zn-Mg alloys (7xxx series)

The Al-Zn-Mg system offers the greatest potential of all aluminium alloys for age hardening although the very high-strength alloys always contain quaternary additions of copper to improve their resistance to stress-corrosion cracking (Section 3.5.5). There is, however, an important range of medium-strength alloys containing little or no copper that have the advantage of being readily weldable. These alloys differ from the other weldable aluminium alloys in that they age harden significantly at room temperature. Moreover, the strength properties that are developed are relatively insensitive to rate of cooling from high temperatures, and they possess a wide temperature range for solution treatment, i.e. $350°$ C and above, with the welding process itself serving this purpose. Thus there is a considerable recovery of strength after welding and tensile strengths of around 320 MPa are obtained without further heat treatment. Yield strengths may be as much as double those obtained for welded components made from the more commonly used Al-Mg and Al-Mg-Si alloys.

Weldable Al-Zn-Mg alloys were first developed for lightweight military bridges but they now have many more commercial applications, particularly in Europe, e.g. Fig. 3.18. Elsewhere, their use has been less widespread for fear of possible stress-corrosion cracking in the region of welds. Many compositions are now available which may contain from 3 to 7 % zinc and 0.8 to 3 % magnesium (Zn + Mg in the range 4.5 to 8.5 %) together with smaller amounts (0.1 to 0.3 %) of one or more of the elements chromium, manganese and zirconium. These elements are added mainly to control grain structure during fabrication and heat treatment although it has been claimed that zirconium also improves weldability. Minor additions of copper are made to some alloys but the amount is kept below 0.3 % to minimize both hot-cracking during the solidification of welds and

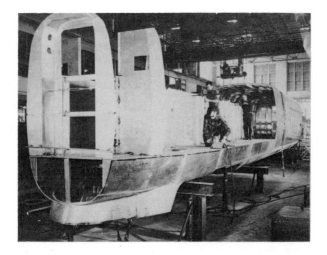

Fig. 3.18   Railway carriage made from a welded Al-Zn-Mg alloy

Table 3.6 Zinc and magnesium contents and ratios in some Al-Zn-Mg and Al-Zn-Mg-Cu alloys

	Alloy	Zn (%)	Mg (%)	Zn + Mg (%)	Zn/Mg ratio
Medium-strength	7104	4.0	0.7	4.7	5.7
weldable	7008	5.0	1.0	6.0	5.0
Al-Zn-Mg	7011	4.7	1.3	6.0	3.7
alloys	7020	4.3	1.2	5.5	3.6
	7005	4.5	1.4	5.9	3.2
	7004	4.2	1.5	5.7	2.8
	7051	3.5	2.1	5.6	1.7
Higher strength	7003	5.8	0.8	6.6	7.2
weldable	7046	7.1	1.3	8.4	5.5
Al-Zn-Mg	7039	4.0	2.8	6.8	1.4
alloys	V92 •	3.3	4.3	7.6	0.77
High-strength	7049	7.7	2.5	10.2	3.1
Al-Zn-Mg-Cu	7050	6.2	2.3	8.5	2.7
alloys	7010	6.2	2.5	8.7	2.5
	7475	5.7	2.3	8.0	2.5
	7001	7.4	3.0	10.4	2.5
	7075	5.6	2.5	8.1	2.2
	7079	4.3	3.3	7.6	1.3

• Russian

corrosion in service. Compositions of representative weldable Al-Zn-Mg alloys are shown in Table 3.6 for different categories of tensile strength.

Improvements in resistance to stress-corrosion cracking have come through control of both composition and heat treatment procedures. With respect to composition, it is well-known that both tensile strength and susceptibility to cracking increase as the Zn + Mg content is raised and it is necessary to seek a compromise when selecting an alloy for a particular application. It is generally accepted that the Zn + Mg content should be less than 6 % in order for a weldable alloy to have a satisfactory resistance to cracking, although the additional controls may be required. More recently it has been proposed by Gruhl that the Zn:Mg ratio is also important and there is experimental evidence which suggests that maximum resistance to stress-corrosion cracking occurs if this ratio is between 2.7 to 2.9 (Fig. 3.19). As shown in Table 3.6, few of the existing commercial alloys do in fact comply with this proposed ratio and this aspect will no doubt receive more attention. Small amounts of copper and, more particularly, silver have been shown to increase resistance to SCC but the addition of silver is considered too great a cost penalty for this range of alloys.

Two changes in heat treatment procedures have led to a marked reduction in susceptibility to stress-corrosion cracking in the weldable alloys. One has been the use of slower quench rates, e.g. air-cooling, from the solution treatment temperature which both minimizes residual stresses and decreases differences in electrode potentials throughout the microstructure. This practice has also had implications with regard to

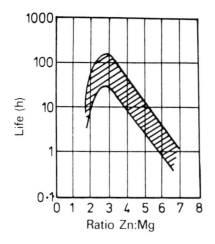

Fig. 3.19 Effect of Zn:Mg ratio on the susceptibility of Al-Zn-Mg alloys to stress-corrosion cracking (from Gruhl, W., *International Congress on Aluminium Alloys in the Aircraft Industry*, Turin, 1976)

composition as there has been a tendency to use 0.08 to 0.25 % zirconium to replace chromium and manganese for the purpose of inhibiting recrystallization because this element has the least affect on quench sensitivity. This characteristic is thought to arise because zirconium forms small, insoluble particles of $ZrAl_3$ whereas chromium and manganese combine with some of the principal alloying elements to form $Al_{12}Mg_2Cr$ and $Al_{20}Cu_2Mn_3$ respectively, thereby removing them from solid solution. The other change has been to artificially age the alloys, sometimes to the extent of using a duplex treatment of the T73 type.

The Al-Zn-Mg alloys are normally welded with an Al-4.5Mg + Mn filler wire although some compositions are available which contain both zinc and magnesium. Although problems with cracking of welded structures in service are now comparatively rare, when this does occur the cracks normally form close to the weld bead/parent alloy interface in what has been termed the white zone (Fig. 3.20). This is a zone within the parent metal which has undergone partial liquation and in which the zinc and magnesium contents vary considerably (Fig. 3.20). It is also one into which elements added to the filler wire can diffuse, at least in part, but the influence of filler composition on cracking has not been studied in detail.

3.5.5 Al-Zn-Mg-Cu alloys (7xxx series)

These alloys have received special attention because it has long been realized that they have the greatest response to age hardening of all aluminium alloys. For example, Rosenhain and his colleagues at the National Physical Laboratory in Britain in 1917 obtained a tensile strength of 580 MPa for a composition Al-20Zn-2.5Cu-0.5Mg-0.5Mn when the value for Duralumin was 420 MPa. However, this alloy and others produced over the next two decades proved to be unsuitable for structural use because of a high susceptibility to stress-corrosion cracking. Because of

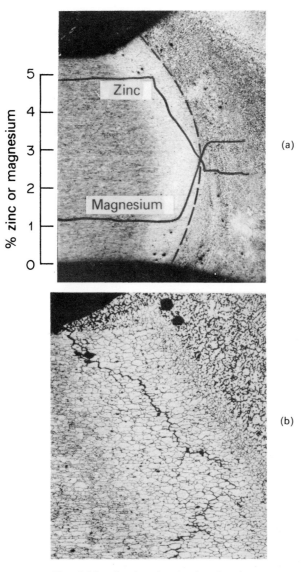

(a)

(b)

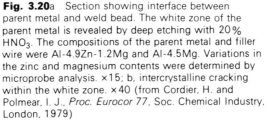

Fig. 3.20a Section showing interface between parent metal and weld bead. The white zone of the parent metal is revealed by deep etching with 20% HNO_3. The compositions of the parent metal and filler wire were Al-4.9Zn-1.2Mg and Al-4.5Mg. Variations in the zinc and magnesium contents were determined by microprobe analysis. ×15; b, intercrystalline cracking within the white zone. ×40 (from Cordier, H. and Polmear, I. J., *Proc. Eurocor 77*, Soc. Chemical Industry, London, 1979)

the critical importance of Al-Zn-Mg-Cu alloys for aircraft construction, this problem has been the subject of continuing research and development and will now be considered in some detail.

Military needs in the late 1930s and 1940s for aircraft alloys having higher strength/weight ratios eventually led to the introduction of several Al-Zn-Mg-Cu alloys of which 7075 is perhaps the best known. Later this alloy and equivalent materials such as DTD 683 in Britain were also accepted for the construction of most civil aircraft. Stronger alloys, e.g. 7178-T6 tensile strength 600 MPa, were introduced for compressively stressed members and another alloy 7079-T6 was developed particularly for large forgings for which its lower quench sensitivity was an advantage. However, continuing problems with stress-corrosion cracking, notably in 7079-T6, and deficiencies in other properties stimulated a need for further improvements. Some aircraft constructors, in fact, reverted to using the lower strength alloys based on the Al-Cu-Mg system even though a significant weight penalty was incurred.

Until then it had been the usual practice to cold-water quench components after solution treatment which could introduce high levels of residual stress. An example is shown in Fig. 3.21 in which machining of the end lug of a cold-water quenched aircraft forging exposed the underlying residual tensile stresses that contributed to stress-corrosion cracking. It is interesting to note that, although cracking within the bore occurred when the forging was in service, the cracks in the sides of the lugs propagated subsequently after the forging had been removed from service and exposed to corrosive atmospheres on separate occasions many years apart. Because

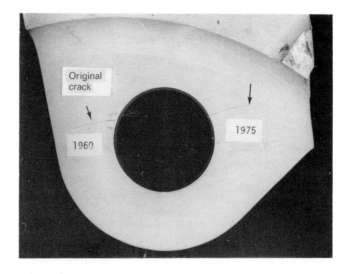

Fig. 3.21   Stress-corrosion cracks in a cold-water quenched Al-Zn-Mg-Cu alloy forging

of the problem of quenching stresses some attempt was made, at least in Britain, to use chromium-free alloys for forgings and other components that could not be stress-relieved. Such alloys could accommodate a milder quench, e.g. in boiling water, without suffering a reduced response to ageing.

Another early practice was to give a single ageing treatment at temperatures in the range 120 to 135°C at which there was a high response to hardening due to precipitation of GP zones (Fig. 3.22). It was known that ageing at a higher temperature of 160–170°C, at which the phase η' (or η) formed did result in a significant increase in resistance to stress-corrosion cracking but tensile properties were much reduced (see curve a in Fig. 3.22). Subsequently, a duplex ageing treatment designated the T73 temper was introduced in which a finer dispersion of the η' (or η) precipitate could be obtained through nucleation from pre-existing GP zones. As shown in Fig. 3.22 (curve b) tensile properties of 7075-T73 are about 15 % below those for the T6 temper but now the resistance to stress-corrosion cracking is greatly improved. For example, tests on specimens loaded in the short transverse direction have shown the alloy 7075 aged to the T73 temper to remain uncracked at stress levels of 300 MPa whereas, in the T6 condition, the same alloy failed at stresses of only 50 MPa in the same environment. Confidence in the T73 temper has been demonstrated by the

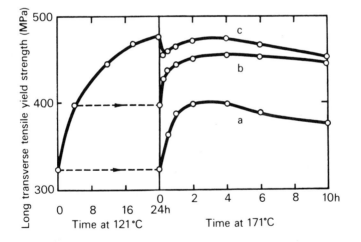

Fig. 3.22 Comparison of yield strengths (0.2% proof stress) of 7075 plate resulting from isothermal (171° C) (curve a) and two-stage (121/171° C) (curve b) precipitation heat treatments. The enhanced yield strength for the alloy 7050 given a two-stage treatment is shown in curve c (from Hunsicker, H. Y., *Rosenhain Centenary Conference on the Contribution of Physical Metallurgy to Engineering Practice*, The Royal Society, London, 1976)

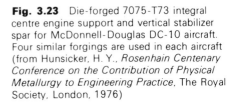

Fig. 3.23   Die-forged 7075-T73 integral centre engine support and vertical stabilizer spar for McDonnell-Douglas DC-10 aircraft. Four similar forgings are used in each aircraft (from Hunsicker, H. Y., *Rosenhain Centenary Conference on the Contribution of Physical Metallurgy to Engineering Practice*, The Royal Society, London, 1976)

use of 7075-T73 for critical aircraft components such as the large die-forging shown in Fig. 3.23.

The use of alloys given the T73 temper has required that some aircraft components be redesigned and weight penalties have resulted when replacing alloys aged to the T6 temper. For this reason, much research has been directed to the development of alloys that could combine a high resistance to stress-corrosion cracking with maximum levels of tensile properties. Some success was achieved with the addition of 0.25 to 0.4% silver as this element modifies the precipitation process in alloys based on the Al-Zn-Mg-Cu system enabling a high response to age hardening to be achieved in a single ageing treatment at 160 to 170°C. One German commercial high-strength alloy designated AZ74 (7009) contains this element. More recently, an alloy 7050 has been developed in the United States in which the level of copper normally present in alloys such as 7075 has been raised in order to increase the strengthening associated with the second stage of the T73 treatment (curve c in Fig. 3.22). Although this alloy presents casting difficulties, it has now been introduced for structural members of several aircraft.

Other compositional changes have been made. One example is a reduction in the level of impurities iron and silicon in alloys such as 7075 which has reduced the size and number of intermetallic compounds that assist crack propagation. The refined alloy is designated 7475 and this change has also improved fracture toughness although a cost penalty of around 10% is involved. Another modification has been to use 0.08 to

0.25 % zirconium as a recrystallization inhibitor in place of chromium or manganese, or in combination with smaller amounts of these elements, in order to reduce quench sensitivity so that slower quench rates can be used (Section 3.5.4), e.g. the alloy 7050 and the British alloy 7010.

Overall, the progress that has been achieved in combating stress-corrosion cracking in the 7xxx series materials through alloy development and changes in heat treatment can be appreciated from results shown in Fig. 3.16b. Contrary to the 2xxx series alloys, improvements have come from large changes to the plateau values for crack growth rate rather than from increases in levels of the threshold stress intensity needed to initiate cracks.

So far as other properties are concerned, duplex ageing treatments have also increased the resistance of the 7xxx alloys to exfoliation (layer) corrosion (T76 temper) although fatigue properties have been little changed (Fig. 3.15). It has been reported that the application of thermomechanical processing to the 7xxx alloys can achieve an improvement of some 20 % either in strength with no loss in toughness, or in toughness at a given stress level. However, commercial exploitation of this form of processing is at an early stage.

3.5.6 Special alloys

Lithium-containing alloys The system Al-Li responds to age hardening due to precipitation of an ordered, metastable phase Al_3Li that is coherent with the aluminium matrix, and which is unlike any other precipitate found in aluminium alloys. Moreover, if lithium is added to the Al-Cu system, this phase forms independently of the normal precipitation sequence in these alloys. A commercial alloy, 2020 (Al-4.5Cu-1.1Li-0.2Cd-0.5Mn), was developed in the United States which showed good creep properties at temperatures as high as 175° C. The alloy also had an elastic modulus some 10 % higher than other aluminium alloys and displayed a high resistance to stress-corrosion cracking. However, ductility and fracture toughness were relatively low and the material has found little commercial use.

More recently, considerable interest has been shown in several Al-Mg-Li alloys in which the precipitate Al_3Li again forms on artificial ageing at around 190° C. Some Russian alloys are available with 5 to 6 % magnesium and 2 % lithium, and other compositions are being investigated having lower magnesium contents. Tensile strengths of around 480 MPa and a satisfactory elongation of 8 % have been reported. Elastic modulus (E) is around 10 % higher than other aluminium alloys which, when combined with the lower density of the Al-Mg-Li alloys ($d \sim 2.5$), means that the specific modulus or stiffness (E/d) may be raised by 20 to 25 %. The alloys offer the prospect of significant weight savings if used for structural members in aircraft, although much work remains to be done to optimize composition and processing so as to improve properties such as fracture toughness.

Superplastic alloys Superplasticity is the ability of certain materials to undergo abnormally large extensions (commonly 1000 % or more) without necking or fracturing. After many years as a laboratory curiosity, the phenomenon has now attracted commercial interest because of the possibility of forming complex shapes in a small number of operations using relatively inexpensive tooling.

The usual requirement for a material to show superplastic characteristics is the presence of roughly equal proportions of two stable phases in a very fine dispersion, e.g. grain size 1 to 2 μm, which must be maintained at the working temperature. Such materials are commonly eutectics or eutectoids and the eutectic alloys Al-33Cu and Al-35Mg exhibit the phenomenon. These alloys have no commercial potential but a new eutectic alloy Al-5Ca-5Zn has been developed in Canada to a stage where a number of prototype products have been produced by superplastic forming at about 450° C. Typical tensile properties at room temperature are 0.2 % proof stress 160 MPa, tensile strength 175 MPa and elongation 12 %, and the alloy has good corrosion resistance.

In Britain attention has been directed to developing a composition having room temperature properties more typical of existing medium-strength aluminium alloys. In this regard it was known that Al-5Cu can recrystallize to give a very fine grain size. However, this structure is not stable at temperatures of 400 to 500° C where superplastic flow might be expected, because most of the CuAl$_2$ particles that might restrict grain growth have redissolved. The possibility of adding a third element that might form small stable particles was then considered, and zirconium was found suitable for this purpose. A composition Al-6Cu-0.5Zr was subsequently developed which was found to be superplastic in the temperature range 420 to 480° C and, under the name of Supral 100, this material is now being used to produce products such as those shown in Fig. 3.24.

The production route for the alloy is basically similar to that used for manufacturing conventional aluminium alloy sheet, although some changes are necessary in order to achieve adequate supersaturation of zirconium. Higher casting temperatures ($\sim$ 800°C) are used and the semi-continuous direct-chill casting process is modified so that the rate of solidification is increased. The cast slabs are sometimes clad with pure aluminium during the hot-rolling stage for subsequent corrosion protection, and reduction to sheet is not greatly different from standard practice. Zirconium precipitates as small particles of a metastable, cubic phase ZrAl$_3$ and the fine grained structure develops during superplastic deformation at temperatures between 420 and 480° C. Strain rates are of the order of 5 x 10^{-3}s^{-1} and, although the flow stresses are greater than those needed to form thermoplastics, similar methods of fabrication can be used. Once a component has been formed, it can be strengthened by precipitation hardening using standard heat treatment operations. Mechanical tests on clad 1.6 mm sheet of the alloy Al-6Cu-0.5Zr after solution treating, quenching and ageing 16 h at 165° C have given values of 0.2 % proof stress 310 MPa, tensile strength 405 MPa, with 9 % elongation.

Fig. 3.24 Examples of products produced by superplastic forming of the alloy Al-6Cu-0.5Zr (courtesy T.I. Superform Ltd)

Similar figures for the alloy in the as-formed condition are 125 MPa, 225 MPa and 15% respectively.

Powder metallurgy products One of the early products produced by powder metallurgy techniques was the material known as SAP that was developed in Switzerland. SAP is an Al-Al_2O_3 alloy prepared by pressing and sintering finely ground flakes of aluminium, and the final microstructure may contain an oxide content as high as 20% in the form of Al_2O_3 particles dispersed in the aluminium matrix. The sintered compacts can be hot-worked by extrusion, forging and rolling to give a product which, using Al-15Al_2O_3 as an example, may have room temperature properties of 0.2% proof stress 240 MPa, tensile strength 345 MPa, with 7% elongation. These properties are relatively low but the fact that the oxide is stable up to the melting point means that creep strength is superior to conventional aluminium alloys at temperatures above about 200 to 250°C. Although SAP was once considered as a possible skin material for high speed aircraft which would suffer aerodynamic heating, it has found relatively few applications and its major use has been for fuel element cans in certain nuclear fission reactors which use organic coolants. For this purpose its creep strength, corrosion resistance, low capture cross-section for neutrons and absence of radioactive isotopes are special advantages. The material is also a candidate for part of the internal structure of experimental nuclear fusion reactors which may employ the deuterium-tritium fuel cycle.

A much wider range of powders is now available including pre-mixes which give compositions similar to several wrought alloys normally produced from cast ingots. Improvements have also been made with lubricants used during pressing the compacts which have reduced the earlier problems of excessive wear of punches and dies. Better combinations of strength and resistance to both corrosion and stress-corrosion cracking have been obtained for small forgings and extrusions made from pre-alloyed, atomized powders than the corresponding products made from ingot metal. In addition, it has been possible to blend into the powders elements that could not be readily added to conventional alloys. One example has been the addition of cobalt to an Al-Zn-Mg-Cu alloy which produces a fine dispersion of the phase CO_2Al_9 in the powder metallurgy product. This phase tends to form along grain boundaries and is claimed to reduce the rate of crack propagation in conditions that normally favour stress-corrosion cracking.

Another technique involving powder metallurgy has been developed with the aim of producing insoluble, submicron particles in aluminium in order to improve strength at elevated temperatures. This involves very rapid solidification which allows a large increase in solute retention of normally sparingly soluble elements such as iron, chromium, and nickel in aluminium. One method of producing these rapidly solidified structures is spray casting in which the melt is atomized with a gas and sprayed onto a moving cold substrate. A strip of porous deposit is obtained which, by subsequent hot-rolling, can be converted to a fully dense material. As much as 5% of elements such as iron and manganese can be retained in supersaturated solid solution which subsequently precipitate as stable compounds during the hot-rolling operation. Unlike precipitates in conventional age-hardened alloys, these particles do not easily coarsen at temperatures in the range 150 to 250°C because of the slow diffusion rates of the elements concerned. Table 3.7 gives an indication of the elevated temperature stability of a spray cast Al-Fe-Ni compact as compared with the conventional alloy 2618 (RR58) that is normally used for elevated temperature applications.

3.6 Joining

Aluminium can be joined by most methods used for other metals including welding, brazing, soldering, bolting, riveting and adhesive bonding.

Table 3.7 Change in tensile strength at room temperature of the spray cast Al-Fe-Ni alloy compact and 2618 (RR58) after exposure for one week at various temperatures

Alloy	Tensile strength at room temperature (MPa) after exposure for one week at		
	Room temperature	150°C	300°C
Al-Fe-Ni compact	560	585	460
2618 (RR58)	465	440	210

Welding and brazing will be considered in some detail and the other methods only referred to briefly.

Mechanical joining by fasteners is covered by a range of engineering codes and poses no real technical problems. Aluminium alloy rivets are normally selected so that their mechanical properties closely match those of the material to be joined. Most commercial rivets are produced from one of the following alloys: 1100, 2017, 2024, 2117, 2219, 5056, 6053, 6061, and 7075. They are usually driven cold and some, e.g. 2024, if solution treated and quenched shortly before use, will gradually harden by ageing at room temperature. Alloy 2024-T4 is commonly used for aluminium bolts or screws although steel fasteners are usually cheaper. Such fasteners must be coated to prevent galvanic corrosion of the aluminium, and plating with nickel, cadmium, or zinc is generally used depending upon the corrosive environment. Under severe conditions, stainless steel fasteners are preferable.

Adhesive bonding is particularly suitable for aluminium because of the minimal need for surface preparation. For general bonding with low-strength adhesives, and in field applications, surfaces may be prepared by mechanical abrasion using wire brushes, abrasive cloths or grit or shot blasting. Where higher bond strengths are required, solvent degreasing is essential and this may need to be followed by some form of chemical treatment. This may involve immersion in an acid bath such as an aqueous solution of sodium dichromate and sulphuric acid, or anodizing which thickens the oxide film and provides an excellent surface for bonding. Adhesives include epoxy and phenolic resins and a range of elastomers.

Soldering involves temperatures below about $425°C$ and tends to be troublesome because aggressive chemical fluxes are needed to remove the oxide film. Problems with corrosion may occur if these fluxes are not completely removed, as well as from galvanic effects that can arise because the low melting point solders have compositions quite different from the aluminium alloys.

3.6.1 Welding

Oxyacetylene welding was widely applied to the joining of aluminium alloys following the development in 1910 of a flux that removed surface oxide film. However, as such fluxes contain halides, any residues can cause serious corrosion problems and this form of fusion welding has been superseded except for a limited amount of joining of thin sheet.

The two main processes in use today are TIG (tungsten inert gas) and MIG (metal inert gas) in which the flux is replaced by a shroud of gas, commonly argon. The heat source is the electric arc struck between either a non-consumable tungsten electrode (TIG) or consumable metal wire (MIG) and the workpiece. Current flows in the arc due to ionization of the inert gas and it is the ionized particles which disrupt the oxide film and clean the surface. TIG welding is mostly carried out manually although the use of mechanized equipment is increasing where higher welding speeds can offset the greater cost of the facility. Where necessary, filler metal is introduced as bare wire. In MIG welding the consumable electrode is fed into the weld pool automatically through a water-cooled gun and the whole process is normally mechanised. It is favoured for volume production work, particularly on material thicker than 4 mm. MIG welding of thin-gauge aluminium presents problems because of the relatively thin wire that would need to be used at normal welding speed which is difficult to handle, or because welding speeds become too high with the thicker wires. For this purpose the pulsed MIG process has been devised in which transfer of metal from the wire tip to the weld pool occurs only at the period of the pulse or peak in welding current. During the intervals between pulses, a background current maintains the arc without metal transfer taking place.

The weldability of the various wrought alloys was mentioned when the individual classes were discussed. Essentially, all are weldable with the exception of most of 2xxx series and the high-strength alloys of the 7xxx series. An example of a welded structure is shown in Fig. 3.18. The strengths of non-heat treatable alloys after welding are similar to their corresponding strengths in the annealed condition irrespective of the degree of cold-working prior to welding. This loss of strength is due to annealing of the zone adjacent to the weld and can only be offset by selecting a weldable alloy such as 5083 which has comparatively high strength in the annealed condition. This accounts for the popularity of this series of alloys for welded constructions. When a heat treatable alloy is welded there is a drop in the original strength to that approximating the T4 condition. The metallurgical condition of this softened zone is complex and may include regions that are partly annealed, solution treated and over-aged, depending upon the actual alloy, speed of welding, material thickness and joint configuration. As in the case of the medium-strength 7xxx alloys (Section 3.5.4), some natural ageing can occur in parts after welding but the overall strength of the welded joint is not usually restored to that of the unwelded alloy. The presence of the range of metallurgical structures adjacent to welds can lead to localized corrosive attack of some alloys in severe conditions.

Filler alloys must be selected with due regard to the composition of the parent alloys and common fillers are listed in Table 3.8. Generally speaking selection is based primarily on ease of producing crack-free welds of the highest strength possible. However, in some cases maximum resistance to corrosion or stress-corrosion, or the ability of the weld to accept a decorative or anodized finish compatible with the parent alloy, may be required.

The fatigue strengths of welded joints tend to show little difference from one alloy to another. Some improvements are possible by mechanical treatments, e.g. peening the weld toes and heat affected zones. One notable advantage of aluminium alloys is that strength and ductility increase as the temperature is reduced and, for this reason, the alloys are particularly suitable for welded cryogenic assemblies.

Table 3.8 Aluminium filler alloys for general purpose TIG and MIG welding of wrought alloys

Base metal welded to base metal	7005	6061 6063 6351	5454 5154A	5086 5083	5052	5005 5050A	3004 Alclad 3004	1100 3003 1200
1100 1200 3003	5356a	4043	4043a	5356b	4043a	4043a	4043a	1100b 1200b
3004 Alclad 3004	5356a	4043a	5356a	5356a	5356ab	4043a	4043a	
5005 5050A	5356a	4043c	5356a	5356a	4043a	4043ad		
5052	5356a	5356ab	5356c	5356a	5356c			
5083 5086	5356a	5356a	5356a	5356a				
5154A	5356a	5356bc	5356ac					
5454	5356a	5356bc	5356ac					
6061 6063 6351	5356a	4043c						
7005	5039a							

Notes:
(a) 5356 or 5556 may be used
(b) 4043 may be used for some applications
(c) 5154A, 5356 and 5556 may be used. In some cases they provide: improved colour match after anodizing treatment; higher weld ductility; higher weld strength; improved stress-corrosion resistance
(d) Filler metal with the same analysis as the base metal is sometimes used.
For alloy compositions see Tables 3.2 and 3.4 except: 4043 Al-5Si; 5356 Al-5Mg-0.1Cr-0.1Mn; 5039 Al-3.8Mg-2.8Zn-0.4Mn-0.15Cr; 5154A Al-3.5Mg-0.3Mn

3.6.2 Brazing

Brazing involves the use of a filler metal having a liquidus above about 425° C but below the solidus of the base metal. Most brazing of aluminium and its alloys is done with aluminium-base fillers at temperatures in the range 560 to 610° C. The non-heat treatable alloys that have been brazed most successfully are the 1xxx and 3xxx series and the low-magnesium members of the 5xxx series. Of the heat treatable alloys, only the 6xxx series can be readily brazed because the solidus temperatures of most of the alloys in the 2xxx and 7xxx series are below 560° C. These alloys can, however, be brazed if they are clad with aluminium or the Al-1Zn alloy 7072. Most commercial filler metals are based on the Al-Si system with silicon contents in the range 7 to 12%.

Removal of the aluminium oxide films is usually considered mandatory and this can be achieved if the surfaces are cleaned and heated to the brazing temperature in the presence of a flux. Such fluxes commonly contain a mixture of alkali and alkaline earth chlorides and fluorides so that subsequent corrosion by residues is possible. Since a major innovation by the aluminium industry has been to replace the copper water-cooling radiator in the motor car by a cheaper and lighter aluminium alloy assembly, much attention has been directed to the possibility of fluxless brazing.

Most processes of this latter type also have as a prerequisite the removal of surface oxide layers. One method involves exposing the assembly to magnesium vapour, followed by brazing in vacuum which requires pumping facilities as well as the use of radiant heating that is relatively inefficient under these conditions. Another uses ultrasonic vibration of the molten brazing alloy in order to remove the oxide film by cavitation. More recently a German process has been developed in which the oxide film is not removed. Instead the wetting capability of the Al-Si brazing fillers has been enhanced by minor additions of antimony, barium, strontium or bismuth. The assembly must first be degreased, lightly etched and dried before brazing, which is carried out in purified nitrogen using thermal convection as the form of heating. This technique has been applied commercially to produce radiators for production vehicles manufactured by the Daimler-Benz Company.

3.7 Special products

3.7.1 Electrical conductor alloys

The use of aluminium and its alloys as electrical conductors has increased significantly in recent decades, due mainly to fluctuations in the price and supply of copper. The conductivity of electrical conductor (EC) grades of aluminium and its alloys average about 62% that of the International Annealed Copper Standard (IACS) but, because of its lower density, aluminium will conduct more than twice as much electricity for an equivalent weight of copper. As a consequence, aluminium is now the least expensive metal with a conductivity high enough for use as an electrical conductor and this situation is unlikely to change in the future.

Aluminium has virtually replaced copper for high voltage overhead transmission lines although the relatively low strength of the EC grades requires that they be reinforced by including a galvanized or aluminium-coated high-tensile steel core with each strand of wire in the cable. Aluminium is also widely used for insulated power cable, especially in underground systems. In this case, instead of the substitution of copper wires by aluminium, each strand of wire is usually replaced by a solid aluminium conductor which is continuously cast by the Properzi process (Section 3.1.1) and sector shaped by rolling (Fig. 3.25). Cable manufacture is thus simplified and economies are also achieved with insulating materials. For other applications, e.g. wiring for electric motors, communication cables or power supply to buildings, properties such as tensile

Fig. 3.25 An underground cable consisting of four-core
solid aluminium conductors; insulated with PVC and
enclosed in lead, then in steel to protect against
mechanical damage, and lastly in PVC as an outer layer

strength and ductility also become critical requirements and growth in the use
of aluminium has been slower.

Stronger alloys such as some heat treatable Al-Mg-Si compositions have
been used as electrical conductors for a number of applications but their
conductivity is relatively low (55 % IACS or less). In general, alloying
elements which are either in solid solution or present as finely dispersed
precipitates cause significant increases in resistivity so that these methods
of strengthening tend to be unacceptable for aluminium conductors. Work
hardening is less deleterious in this regard, but conductors strengthened in
this way tend to exhibit poor thermal stability and may be susceptible to
mechanical failure in service. For these reasons, attention has been directed
to alternative methods by which adequate strengthening can be achieved
whilst retaining a high electrical conductivity (>60 % IACS).

As mentioned in Section 3.3.2, aluminium is amenable to substructure
strengthening and much of the developmental work has been directed at
stabilizing the substructure with low volume fractions of finely dispersed
intermetallic compounds. These compounds also assist in improving
ductility in the final product by increasing strain hardening which delays
localized deformation and necking. It is necessary for the compounds to be
uniformly distributed and this presents some difficulties because they
precipitate in the interdendritic regions during casting. Casting methods
must be used which ensure a rapid rate of solidification as this reduces the
dendrite arm spacing and refines the microstructure. Extensive deformation during rod production and wire drawing further assists in distribut-
ing the compounds. Rod production involves hot-working at a temperature at which dynamic recovery occurs during processing, thereby ensuring

the formation of subgrains rather than cells, and the work-hardening exponent *n* approaches 0.5.

The compositions and properties of some new conductor alloys are compared with those of the older EC materials in Table 3.9. Additions such as iron and nickel have been selected because they have a very low solubility in aluminium, they form stable compounds and they cause relatively small increases in resistivity when out of solution. The presence of magnesium in the alloy 8076 leads to improved creep resistance. The success of the new products has been reflected in their increased usage for electrical wire which, in the United States, doubled in the decade from 1965 to 1975.

Table 3.9 Compositions and typical properties of some aluminium alloy electrical conductor wires (from Starke, E. A. Jr., *Mater. Sci. Eng.*, **29**, 99, 1977)

	Alloy	Yield strength (MPa)	Tensile strength (MPa)	Elongation in 250 mm (%)	Electrical conductivity (% IACS)
Old alloys	EC (99.6 Al)	28	83	23	63.4
	5005-HI9 (Al-0.8Mg)	193	200	2	53.5
	6201-T81 (Al-0.75Mg-0.7Si)	303	317	3	53.3
New alloys	Triple E (Al-0.55Fe)	68	95	33	62.5
	Super T (Al-0.5Fe-0.5Co)	109	129	25	61.1
	8076 (Al-0.75Fe-0.15Mg)	61	109	22	61.5
	Stabiloy (Al-0.6Fe-0.22Cu)	54	114	20	61.8
	Nico (Al-0.5Ni-0.3Co)	68	109	26	61.3
	X8130 (Al-0.6Fe-0.08Cu)	61	102	31	62.1

Note: All alloys except 5005 and 6201 are in annealed condition (0-temper)

3.7.2 Aluminium alloy bearings

The development of aluminium alloys for bearings dates back to the 1930s when the high thermal conductivity, corrosion resistance and fatigue resistance of aluminium were recognized. Compositions containing 2 to 15% copper were then the most successful and the structure of the alloys comprised hard intermetallic compounds in a softer aluminium matrix. Although they were adopted for a number of applications, their relatively high hardness was a disadvantage with regard to certain property requirements, e.g. conformability to rotating shafts. Al-Sn alloys offered the alternative prospect of a softer bearing alloy and compositions with up to 7% tin were introduced. These are still used. They were first produced by casting solid bearings but the high coefficient of thermal expansion of aluminium made retention of fit with steel shafts virtually impossible at the operating temperatures of modern engines. Consequently, most current aluminium alloy bearings are backed with steel.

Early work revealed that the seizure resistance of Al-Sn bearings continued to improve as the tin content was raised to levels of 20% or more. However, the full potential of these alloys could not then be realized because, regardless of the method of casting, tin contents in excess of 7 % resulted in the formation of low-strength, grain boundary films of tin (Fig. 3.26). Eventually it was discovered that these films could be dispersed into discrete globules if the alloys were cold-worked and recrystallized during processing. Bonding the bearing alloy to a steel backing presented a further problem because intimate contact was prevented by the tenacious oxide film on the alloy until it was found that this film could be fragmented and dispersed by severe cold-rolling.

Current practice for producing steel-backed, Al-Sn bearings commonly involves the following procedures:

(i) Chill casting the alloy on a copper plate to promote directional solidification of a slab approximately 25 mm thick.

(ii) Cladding the sides of the slab with pure aluminium and heavily cold-rolling to produce a bonded composite having a thickness less than one-thirtieth that of the original size. This treatment elongates the tin particles.

(iii) Cold roll-bonding the composite to the steel backing, contact being made with the pure aluminium surface. The configuration of the rolls is arranged so that the bearing alloy is more heavily cold-worked than the steel backing so that the latter is not unduly hardened.

(iv) Annealing at 350°C for about 1 h which recrystallizes the bearing alloy and coalesces the tin into discrete globules in grain corners in what is known as reticular structure (Fig. 3.26).

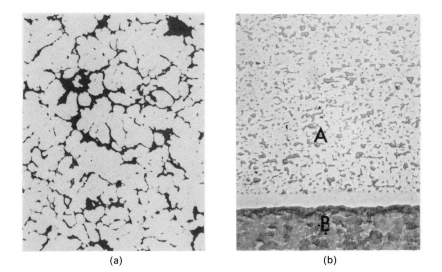

(a) (b)

Fig. 3.26a Continuous grain boundary film of tin in a cast Al-30Sn-3Cu alloy (from Liddiard, E. A. G., *The Engineers' Digest,* 1955). ×120; b, microstructure of Al-20Sn-1Cu (A) roll-bonded to steel (B) via an aluminium interlayer (courtesy G. C. Pratt, The Glacier Metal Co Ltd). ×150

A content of 20 % tin is commonly regarded as an optimal level although some commercial bearings contain as much as 30 % tin. In addition, 1 % copper is often added to increase the strength and improve the properties of the aluminium matrix. These alloys offer the further advantage of not requiring the surface overlay that is needed to protect copper-lead and some other bearings from corrosion by chemicals present in modern lubricating oils. Steel backed Al-Sn alloys are widely used for diesel engines and in many European and North American motor cars, particularly for crankshaft bearings.

Two other aluminium alloys have been developed as bearing materials. One is based on the Al-Pb system in which the cheapness of lead is an advantage. Gross segregation is a problem in casting these materials due to the immiscibility of aluminium and lead in both the liquid and solid states, and to their large differences in density. However, a successful process has been developed to produce strip by rolling pre-mixed powders. The second material is Al-Si which offers greater fatigue strength than Al-Sn alloys and it is used in some high speed diesel engines.

3.7.3 Automotive alloys

The use of Al-Sn bearing alloys and aluminium radiators has been discussed above and cast aluminium alloys for engines are considered in Section 4.2.3. Elsewhere the use of aluminium alloys in motor cars has been limited because they are more expensive and difficult to process than steels. Now the escalating cost of fuel has stimulated a renewed interest in the possibility of achieving weight savings by the use of aluminium alloys for body sheet, bumpers and other components. Many predictions have been made concerning the future use of aluminium. For example it has been proposed that a typical 1977 model car in the United States that weighed 1600 kg and contained 45 kg of aluminium may, for the same size in 1985, weigh 1135 kg and contain 225 kg of aluminium. Nevertheless it should be remembered that to produce 1 kg of aluminium requires more than six times as much energy than is needed to produce 1 kg of steel. Moreover, aluminium is in competition with plastics as well as with improved mild steels (HSLA steels) containing microadditions of elements such as niobium which have higher strength:weight ratios. Fig. 3.27 summarizes a recent analysis of relationships of costs and weight reductions in replacing mild steel body panels (MSBP) with aluminium sheet and other materials.

Depending upon the application, aluminium alloys for body sheet must have high strength with adequate formability, be dent and corrosion resistant, and have good surface appearance, e.g. free of Lüders bands. Formability, notably the capacity of sheet to be drawn, is controlled by crystallographic texture developed on rolling as well as by the work-hardening component n of the alloy and R-value (ratio of width to thickness strain), both of which should be as large as possible (Section 3.3). An optimal combination of formability and strength requires careful control of alloy composition and heat treatment.

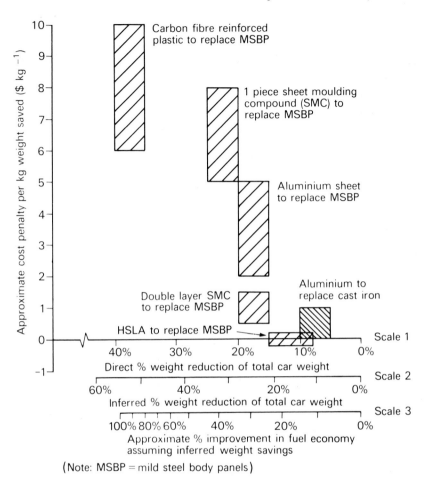

Fig. 3.27 Approximate relationship of costs (expressed in US dollars) and weight reductions in replacing mild steel body panels (MSBP) with aluminium sheet, high-strength low alloy (HSLA) steels, plastic sheet moulding compounds (SMC) or carbon fibre reinforced plastics. Inferred weight reductions include direct or indirect weight savings (courtesy R. C. Schodde, BHP Melbourne Research Laboratories, 1979)

The first alloys selected for automotive sheet were 3004, 5052 and 6061. However, the low strength of 3004, the appearance of Lüders lines during the drawing of 5052 and the limited formability of 6061, led to the development of new compositions, e.g. 2036 (Al-2.6Cu-0.45Mg-0.25Mn), 5182 (Al-4.5Mg-0.35Mn), 6009 (Al-1.1Si-0.6Mg-0.35Mn) and 6010 (Al-1Si-0.8Mg-0.5Mn-0.4Cu). In some alloys, the levels of impurities are reduced to improve formability. Those alloys that respond to heat

Table 3.10 Typical mechanical properties of some body sheet alloys (from Starke, E. A. Jr, *Mater. Sci. Eng.*, **29**, 99, 1977)

Alloy and temper	0.2% proof stress (MPa)	Tensile strength (MPa)	Elongation (%)	n	R-value
2036-T4	195	340	24	0.23	0.75
5182-0	130	275	21	0.33	0.80
6009-T4	125	230	25	0.23	0.70
6009-T6	325	345	12	–	–
6010-T4	180	290	24	0.22	0.70
6010-T6	370	385	11	–	–
6151-T4	165	260	23	0.19	0.55

treatment can be strengthened by age hardening during the paint baking operation. Mechanical properties of body sheet alloys are given in Table 3.10. The stronger alloys are suitable for outer panels whereas the lower strength but more formable alloys are preferred for complex inner panels. Nevertheless, it is desirable for combinations of alloys in the same assembly to have compositions that are fairly similar, e.g. 6009 and 6010, otherwise the need for scrap separation imposes an increase in costs.

Higher strength alloys are required for bumpers and bumper reinforcements and attention has been directed to the 7xxx series. Both Al-Zn-Mg and Al-Zn-Mg-Cu alloys have been developed, the latter being derived from compositions traditionally used for aircraft construction. An example is X7029-T5 (Al-4.7Zn-1.6Mg-0.7Cu) which has the following mechanical properties: 0.2% proof stress 380 MPa, tensile strength 430 MPa and elongation 15%. Impurity levels are kept low to improve toughness and bright finishing characteristics.

3.7.4 Packaging

During the last decade, the use of aluminium in the packaging industry has increased to such an extent that it is now the second largest market for the metal in many countries, including the United States. This situation has come about mainly because aluminium is an attractive material for packaging food and beverages since it has good corrosion resistance, is physiologically harmless, provides good thermal conduction and is impenetrable by light, oxygen, moisture and micro-organisms. Moreover, simple alloys can be used which are easy to recycle. The largest production items are cans and foil.

The entry of aluminium into the can market occurred in 1962 with the introduction of the tear-top tab or easy-open end which was used as the lid for steel cans. Then followed the two-piece, all aluminium can in which the seamless can is produced from sheet by cupping followed by drawing and ironing the side walls. One stage in the combined drawing and ironing process is shown schematically in Fig. 3.28 and it can be seen that ironing

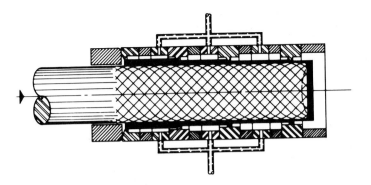

Fig. 3.28 One stage in the drawing and ironing of the seamless aluminium can

involves progressive squeezing of the metal between an inner punch and outer dies having decreasing internal diameters.

Extreme formability with minimum anisotropy and good mechanical strength is required of the sheet. The alloys best meeting these requirements are based on the non-heat treatable 3xxx and 5xxx series. Established alloys such as 3003 and 5086 were used initially but problems in forming thin gauges led to a number of minor but important compositional changes. For example, 3003 (Al-1.2Mn) was strengthened by adding 1% magnesium (3004) and, more recently, an alloy 3005 with lower magnesium content has been developed having properties intermediate between the other two. Several changes have been made to 5086 (Al-4Mg-0.5Mn-0.15Cr). For example, the newer alloy 5082 has more magnesium (4.5%), and less manganese (0.15% maximum) and lower amounts of the impurity elements. There has been a trend to reduce the levels of elements forming intermetallic compounds and inclusions have been minimized by increased attention to filtering the molten aluminium during casting. This step is also of critical importance in the production of alloys for foil.

Foil is commonly defined as rolled sheet having a thickness less than 0.15 mm. Products range from light gauge foil, e.g. 0.01 mm thick, for domestic and other uses, which is made from various grades of commercial-purity aluminium, to thicker foil, for semi-rigid containers, that is usually made from alloys such as 3003 or 8011. Control of impurity contents and the size of intermetallic compounds is again critical to avoid tearing during foil rolling.

The major advance in foil production has been the development of methods for rolling at high speeds whilst maintaining close control of thickness. Rolls capable of handling coil stock weighing as much as 10 tonnes and operating at speeds in excess of 40 m s^{-1} are now common. The coil stock is produced by rolling semi-continuously cast ingots (Fig. 3.1) or, more economically, continuously cast strip (Fig. 3.2b) and then cold-rolling to a strip thickness of about 0.5 mm. Foil is produced by further cold-rolling the strip in several stages in a continuous mill, the material being pack-rolled double in the last stage if very light gauges, e.g. <0.025 mm, are needed. Most foil is then softened by annealing. One of a number of coating treatments such as laminating with paper, lacquering, printing and embossing may then be applied for purposes such as advertising.

3.7.5 Composites

As service requirements have become more demanding, increasing attention has been paid to composites which exhibit properties superior to those of the component materials. Laminates and fibre-reinforced materials are the most common examples.

Sandwich panels are one form of laminate that deserve special mention because they offer the desirable combination of high rigidity and low weight. Such panels comprise thin facings which are secured, usually by adhesive bonding, to a relatively thick, low-density core material. From the design viewpoint, a sandwich panel is similar to an *I*-beam with the facings and core corresponding to the flanges and centre web respectively. The facings carry axial compressive and tensile stresses whereas the core sustains shear and prevents buckling of the facings under compressive loading.

Moment of inertia is a direct measure of stiffness or rigidity and it is interesting to compare values for a sandwich panel and a homogeneous isotropic plate of the same material as the facings of the panel. For example, a sandwich panel with two 0.5 mm thick facings of an aluminium alloy and a balsa wood core 6 mm thick would weigh 3.4 kg m^{-2}. A plate of the same alloy of the same size and weight would be 1.25 mm thick. The moment of inertia of the cross-section of the plate I_p about its neutral axis, per unit width, is given by:

$$I_p = \frac{t_p{}^3}{12}$$

where t_p = thickness of the plate. Neglecting the small effect of the core material, the moment of inertia of the sandwich panel I_s, per unit width, is given by:

$$I_s = 2t_f \left(\frac{t_f + t_c}{2} \right)^2$$

where t_f = thickness of the facings, t_c = thickness of the core. Therefore, for the sandwich panel and plate under consideration, $I_s = 10.5$ mm^4 and $I_p = 0.162$ mm^4. Thus the sandwich panel has the advantage of 65 times the rigidity of a plate having the same weight. It can also be shown that, for the aluminium alloy plate to have equal rigidity, it would weigh four times more than the sandwich panel.

Sandwich panels with cores of balsa wood or foamed plastic are now

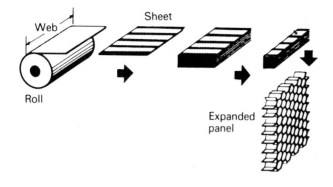

Fig. 3.29 Expansion process for the manufacture of aluminium honeycomb cores for sandwich panels (courtesy Hexel Aerospace)

used for applications such as siding for refrigerated trucks and for a variety of aircraft components. In this latter regard, even greater weight savings are possible by using a honeycomb core made from impregnated paper or aluminium alloy foil. The honeycomb is usually made by an expansion method (Fig. 3.29) which begins with stacking of sheets of foil on which adhesive stripes have been printed. The adhesive is cured and the block cut into slices which are then expanded to form the honeycomb panel. Honeycomb sandwich panels are used in aircraft for applications such as fuselage and wing panels, an example being leading edge flaps, a section of which is shown in Fig. 3.30.

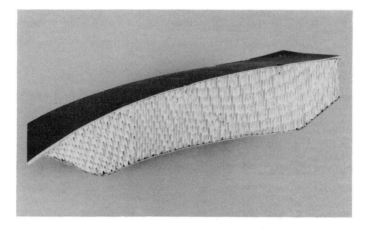

Fig. 3.30 Section of a honeycomb sandwich panel from an aircraft wing flap

Fibre-glass is the most widely known fibre-reinforced composite but its use at elevated temperatures is restricted because the polymeric matrix degrades. Replacing polymers with metals would offer a marked improvement in this regard and many attempts have been made to incorporate strong wires and other fibres in a matrix of aluminium. Examples are hard-drawn stainless steel wires, silica fibres and, more recently, stiff fibres of boron ($E/d = 183$ as compared with 26 for aluminium alloys). Various composite components have been made by interleaving aluminium foils and boron fibres and then compacting by slow hot-pressing. The tendency for boron to be degraded by reacting with aluminium is much reduced by first coating the fibres with silicon carbide producing so-called 'borsic' fibres. Such composites can be operated at temperatures around 320° C and are being investigated as possible materials for compressor blades in gas turbine engines, as well as for certain structural components. Boron fibres have also been incorporated in experimental composites having titanium or magnesium matrices.

Further reading

Van Horn, K. R. (ed), *Aluminum*, Volumes 1–3, American Society for Metals, Cleveland, 1967

Altenpohl, D., *Aluminium and Aluminium Alloys*, Springer-Verlag, Berlin, 1965 (in German)

Van Lancker, M., *Metallurgy of Aluminium Alloys*, Chapman and Hall, London, 1967

Varley, P. C., *The Technology of Aluminium and its Alloys*, Newnes-Butterworths, London, 1970

Mondolfo, L. F., *Aluminium Alloys: Structure and Properties*, Butterworths, London, 1976

Mondolfo, L. F., *Manganese in Aluminium Alloys*, The Manganese Centre, Neuilly sur Seine

The Properties of Aluminium and its Alloys, Aluminium Federation, Birmingham, 1973

Source Book on Selection and Fabrication of Aluminum Alloys, American Society for Metals, Cleveland, 1978

Proc. of Int. Conf. Aluminium and Automobile, Aluminium-Zentrale, Düsseldorf, 1976

Emley, E. F., Continuous casting of aluminium, *Int. Met. Rev.*, **21**, 75, 1976

Talbot, D. E. J., Effects of hydrogen in aluminium, magnesium, copper and their alloys, *Int. Met. Rev.*, **20**, 166, 1975

Bryant, A. J., Metallurgical factors affecting the control of mechanical properties in high-strength aluminium alloys, *Aluminio E. Nuova Metallurgia*, **42**, 228, 1977

Registration Board of International Alloy Designations and Chemical Composition Limits for Wrought Aluminum and Wrought Aluminum Alloys, Aluminum Association, New York, 1976

Tempers for Aluminum Alloy Products, Aluminum Association, New York, 1976

Staley, J. T., Aluminium alloy and process developments for aerospace, *Met. Eng. Quarterly*, **16**, No. 5, 52, 1976

Starke, E. A. Jr., Aluminium alloys for the 70s, *Mater. Sci. Eng.*, **29**, 99, 1977

Kent, K. G., Weldable Al-Zn-Mg alloys, *Metals and Materials*, **4**, 135, 1970

Speidel, M. O., Stress-corrosion cracking of aluminium alloys, *Met. Trans.*, **6A**, 631, 1975

Hunsicker, H. Y., Development of Al-Zn-Mg-Cu alloys for aircraft, *Rosenhain Centenary Conference on the Contribution of Physical Metallurgy to Engineering Practice*, The Royal Society, London, 1976; p. 245

Sanders, T. H. Jr. and Staley, J. T., In: *Fatigue and Microstructure*, American Society for Metals, Cleveland, 1979; Chapter 14

Forsyth, P. J. E. and Stubbington, C. A., Directionality in structure/property relationships: aluminium and titanium, *Met. Technology*, **2**, 158, 1975

Watts, B. M., Superplasticity in Al-Cu-Zr alloys, *Met. Sci.*, **10**, 189, 1976

Schultze, W. and Schoer, H., Fluxless brazing of aluminium using protective gas, *Welding J.*, **52**, 644, 1973

Klock, H. and Schoer, H., *Aluminium Processing by Welding and Brazing*, Deutscher Verlag für Schweissechnik, Düsseldorf, 1977 (in German)

Kerth, W. *et al.*, Aluminium foil production, *Int. Met. Rev.*, **20**, 185, 1975

Mastrovich, J. D., Aluminium can manufacture, *Caltex Lubrication*, **30**, 73, 1975

Pratt, G. C., Materials for plain bearings, *Int. Met. Rev.*, **18**, 62, 1973

Aluminium Alloys in the Aircraft Industries, Technicopy Limited, Stonehouse, 1978

4

Cast aluminium alloys

Aluminium is one of the most versatile of the common foundry metals, with cast products consuming, as a world average, some 20% of this metal. Examples of recent figures are: the United States 15%, Britain and West Germany 23%, Japan 27%, France 29% and Italy 37%. These differences arise mainly because of the greater usage of aluminium alloy castings for transport applications in Europe and Japan (60 to 75% of all aluminium castings, respectively) as compared with the United States (44%).

The most commonly used processes are sand casting, permanent mould (gravity die) casting and cold chamber pressure diecasting. In the United States, for example, these three processes accounted for 90% of all aluminium alloy castings in 1976, the division between each being sand castings 12.5%, permanent mould castings 21% and pressure diecastings 66.5%. Sand moulds are fed with molten metal by gravity. The metal moulds used in permanent mould casting are fed either by gravity or by using low-pressure air or other gas to force the metal up the sprue and into the mould. In pressure diecasting, molten aluminium is forced into a steel die at high pressure by the action of a hydraulic ram. General features about the melting and alloying of aluminium as well as grain refinement, where needed, are similar to those described in Chapter 3.

Apart from light weight, the special advantages of aluminium alloys for castings are the relatively low melting temperatures, negligible solubility for all gases except hydrogen, and the good surface finish that is usually achieved with final products. Most alloys also display good fluidity and compositions can be selected with solidification ranges appropriate to particular applications. The major problem with aluminium castings is the relatively high shrinkage of between 3.5 and 8.5% that occurs during solidification. Allowance for this must be made in mould design in order to achieve dimensional accuracy and to avoid problems such as hot tearing or cracking, residual stresses and shrinkage porosity.

As with wrought materials, there are cast alloys which respond to heat treatment and these are discussed below. It should be noted, however, that pressure diecastings are not normally solution treated because blistering may occur due to the expansion of air entrapped during the casting process. Moreover, there is the possibility of distortion as residual stresses are relieved.

In all areas, except creep, castings normally have mechanical properties that are inferior to wrought products and these properties also tend to be

110

much more variable throughout a given component. As it is common practice to check tensile properties by casting separate test bars, it is necessary to bear in mind that the results so obtained should be taken only as a guide. The actual properties of even a simple casting may be 20 to 25 % lower than the figures given by the test bar.

The demand for a greater assurance in meeting a specified level of mechanical properties within actual castings has led to the concept of 'premium quality' castings and represents a major advance in foundry technology. Specifications for such castings require that guaranteed minimum property levels are met in any part of the casting and match more closely those obtained in test bars. Mechanical properties previously thought unattainable have been achieved through strict control of factors such as melting and pouring practices, impurity levels, grain size and, in the case of sand castings, the use of metal chills to increase solidification rates. Certain radiographic requirements may also need to be met. Moreover, castings are submitted in lots and if one casting selected at random fails to meet specification requirements, all are rejected. Premium quality castings are more expensive to produce although they may be cost-effective if wrought components can be replaced. Some of the improved procedures have been translated into more general foundry practices.

4.1 Designation, temper and characteristics of cast aluminium alloys

No internationally accepted system of nomenclature has so far been adopted for identifying aluminium casting alloys. However, the Aluminum Association of the United States has recently introduced a revised system which has some similarity to that adopted for wrought alloys and this is described below. Details are also given of the current British system of identification.

4.1.1 United States Aluminum Association system

This Association now uses a four digit numerical system to identify aluminium and aluminium alloys in the form of castings and foundry ingot. The first digit indicates the alloy group as follows:

	Current designation	Former designation
Aluminium, 99.00 % or greater	1xx.x	
Aluminium alloys grouped by major alloying elements:		
Copper	2xx.x	1xx
Silicon with added copper and/or magnesium	3xx.x	3xx
Silicon	4xx.x	1 to 99
Magnesium	5xx.x	2xx
Zinc	7xx.x	6xx
Tin	8xx.x	7xx
Other element	9xx.x	
Unused series	6xx.x	

In the 1xx.x group, the second two digits indicate the minimum percentage of aluminium, e.g. 150.x indicates a composition containing a minimum of 99.50 % aluminium. The last digit, which is to the right of the decimal point, indicates the product form with 0 and 1 being used to denote castings and ingot respectively.

In the 2xx.x to 9xx.x alloy groups, the second two digits have no individual significance but serve as a number to identify the different aluminium alloys in the group. The last digit, which is to the right of the decimal point, again indicates product form. A modification to the above grouping of alloys is used in Australia in which the 6xx.x series is allocated to Al-Si-Mg alloys.

When there is a modification to an original alloy, or to the normal impurity limits, a serial letter is included before the numerical designation. These letters are assigned in alphabetical sequence starting with A but omitting I, O, Q and X, the X being reserved for experimental alloys. The temper designations for castings are the same as those used for wrought products (see Fig. 3.10). This does not apply for ingots.

4.1.2 British system

Most alloys are covered by the British Standard 1490 and compositions for ingots and castings are numbered in no special sequence and have the prefix LM. The condition of castings is indicated by the following suffixes:

M   – as-cast
TB   – solution treated and naturally aged (formerly designated W)
TB 7 – solution treated and stabilized
TE   – artificially aged after casting (formerly P)
TF   – solution treated and artificially aged (formerly WP)
TF 7 – solution treated, artificially aged and stabilized (formerly WP-special)
TS   – thermally stress-relieved.

The absence of a suffix indicates that the alloy is in ingot form.

There are also some aerospace alloys which are covered separately by the L series of British Standards and by the DTD specifications. These specifications, which are not systematically grouped or numbered, are concerned with specific compositions and often with particular applications. In both cases, the condition of the casting is not indicated by a suffix and is identified only by the number of the alloy.

4.1.3 General characteristics

The features of the different types of aluminium casting alloys are discussed in succeeding sections and are classified according to the major alloying element that is present, e.g. Al-Si, Al-Mg etc. Because of the problems of identification in different countries, individual alloys are represented by their basic compositions. It will be noted that castings in commercial-purity aluminium are not considered separately because their only major use is for certain electrical applications which were discussed in Chapter 3.

Although many different casting alloys are available, the number used in large quantities is much smaller. In Britain, for example, most cast aluminium alloy components are made from only four alloys designated LM2, LM4, LM6, and LM21. A representative list of alloy compositions is given in Table 4.1 together with the type of casting process for which they are used. Some mechanical properties and the relative ratings of the various casting characteristics are given in Table 4.2.

Selection of alloys for castings which are produced by the various casting processes depends primarily upon composition which, in turn, controls characteristics such as solidification range, fluidity, susceptibility to hot-cracking, etc. Sand castings impose the least limitation on choice of alloy and commonly used alloys are 208 (Al-4Cu-3Si), 413 (Al-11.5Si), 213 (Al-7Cu-2Si-2.5Zn) and 356 (Al-7Si-0.3Mg). Alloys 332 (Al-9Si-3Cu-1Mg) and 319 (Al-6Si-4Cu) are favoured for permanent mould castings and 380 (Al-8.5Si-3.5Cu) and 413 (Al-11.5Si) are most commonly used for pressure diecastings. In the latter case, the prime consideration is a low melting point which increases production rates and minimizes die wear.

In order of decreasing castability, the groups of alloys can be classified in the order 3xx, 4xx, 5xx, 2xx, and 7xx. Corrosion resistance is also a function of composition and the copper-free alloys are generally regarded as having greater resistance than those containing copper. The 8xx series is confined to Al-Sn bearing alloys which were discussed in Section 3.7.2.

4.2 Alloys based on the aluminium-silicon system

Alloys with silicon as the major alloying addition are the most important of the aluminium casting alloys mainly because of the high fluidity imparted by the presence of relatively large volumes of the Al-Si eutectic. Other advantages of these castings are high resistance to corrosion, good weldability and the fact that silicon reduces the coefficient of thermal expansion. However, machining may present difficulties because of the presence of hard silicon particles in the microstructure. Commercial alloys are available with hypoeutectic, eutectic and, less commonly, hypereutectic compositions.

The eutectic is formed between an aluminium solid solution containing just over 1 % silicon and virtually pure silicon as the second phase. The eutectic composition has been a matter of debate but it is now generally accepted as being close to Al-12.7 %Si. Slow solidification of a pure Al-Si alloy produces a very coarse microstructure in which the eutectic comprises large plates or needles of silicon in a continuous aluminium matrix (Fig. 4.1a). The eutectic itself is composed of individual cells within which the silicon particles appear to be interconnected. Alloys having this coarse eutectic exhibit low ductility because of the brittle nature of the large silicon plates. Rapid cooling, as occurs during permanent mould casting, greatly refines the microstructure and the silicon phase assumes a fibrous form with

Table 4.1 Compositions of selected aluminium casting alloys

Aluminum Association number	BS 1490 LM number	Casting process	Si	Fe	Cu	Mn	Mg	Cr	Ni	Zn	Ti	Other
150.1	LM 1	Ingot	*	*	0.10					0.05		99.5 Al min
201.0		S	0.10	0.15	4.0–5.2	0.20–0.50	0.15–0.55				0.15–0.35	Ag 0.40–1.0
208.0		S	2.5–3.5	1.2	3.5–4.5	0.50	0.10		0.35	1.0	0.25	
213.0		PM	1.0–3.0	1.2	6.0–8.0	0.6	0.10		0.35	2.5	0.25	
238.0	LM 4	S and PM	4.0–6.0	0.8	2.0–4.0	0.20–0.6	0.15		0.30	0.50	0.20	
242.0		PM	3.5–4.5	1.5	9.0–11.0	0.6	0.15–3.5		1.0	1.5	0.25	
295.0	LM 14	S and PM	0.7	1.0	3.5–4.5	0.35	0.15–0.35	0.25	1.7–2.3	0.35	0.25	
308.0		S	0.7–1.5	1.0	4.0–5.0	0.35	0.03			0.35	0.25	
319.0		PM	5.0–6.0	1.0	4.0–5.0	0.50	0.10			1.0	0.25	
328.0	LM 21	S and PM	5.5–6.5	1.0	3.0–4.0	0.20–0.6	0.10	0.35	0.25	1.0	0.25	
A332.0		PM	7.5–8.5	1.2	1.0–2.0	0.35	0.20–0.6		2.0–3.0	1.5	0.25	
355.0	LM 13	S and PM	11.0–13.0	0.6•	0.50–1.5	0.50•	0.7–1.3			0.35	0.25	
356.0	LM 16	S and PM	4.5–5.5	0.6	1.0–1.5	0.35	0.40–0.6			0.35	0.25	
A356.0	LM 29	S and PM	6.5–7.5	0.6	0.25	0.35	0.20–0.40			0.35	0.25	
360.0	LM 25	S and PM	6.5–7.5	0.20	0.20	0.10	0.20–0.40			0.10	0.20	
380.0	LM 9	D	9.0–10.0	2.0	0.6	0.35	0.40–0.6		0.50	0.50		
A380.0	LM 24	D	7.5–9.5	2.0	3.0–4.0	0.50	0.10		0.50	3.0		
390.0	LM 24	D	7.5–9.5	1.3	3.0–4.0	0.50	0.10		0.50	3.0		
	LM 30	D	16.0–18.0	1.3	4.0–5.0	0.10	0.45–0.65			0.10		
	LM 6	S, PM and D	10.0–13.0	0.6	0.10	0.50	0.10		0.10	0.10	0.20	

Aluminum Association number	BS 1490 LM number	Casting process	Si	Fe	Cu	Mn	Mg	Cr	Ni	Zn	Ti	Other
413.0	LM 20	D	11.0–13.0	2.0	1.0	0.35	0.10		0.50	0.50	0.20	
	LM 2	D	9.0–11.5	1.0	0.7–2.5	0.50	0.30		0.50	2.0	0.25	
443.0	LM 18	S	4.5–6.5	0.8	0.6	0.50	0.05	0.25		0.50	0.25	
514.0	LM 5	S	0.35	0.50	0.15	0.35	3.5–4.5			0.15		
518.0		D	0.35	1.8	0.25	0.35	7.5–8.5		0.15	0.15		
520.0	LM 10	S	0.25	0.30	0.25	0.15	9.5–10.6				0.25	
535.0		S	0.15	0.15	0.05	0.1–0.25	6.2–7.5				0.10–0.35	
705.0		S and PM	0.20	0.8	0.20	0.40–0.6	1.4–1.8	0.20–0.40		2.7–3.3	0.25	
707.0		S and PM	0.20	0.6	0.20	0.40–0.6	1.4–1.8	0.20–0.40		4.0–4.5	0.25	
712.0		PM	0.15	0.50	0.25	0.10	0.50–0.65	0.40–0.6		5.0–6.5	0.10–0.25	
713.0		S and PM	0.25	1.18	0.40–1.0	0.6	0.20–0.50	0.35	0.15	7.0–8.0	0.25	
850.0		S and P	0.7	0.7	0.7–1.3	0.10	0.10		0.7–1.3		0.20	Sn 5.5–7.0

Notes: Compositions are in % maximum by weight unless shown as a range

S = sand casting; PM = permanent mould (gravity die) casting; D = pressure diecasting

* Ratio Fe:Si minimum of 2:1

● If iron exceeds 0.45% manganese content must be less than one-half the iron content

Table 4.2 Typical mechanical properties and foundry characteristics of selected aluminium casting alloys

Aluminum Association number	BS 1490 LM number	Casting process	Temper	0.2% proof stress (MPa)	Tensile strength (MPa)	Elongation (% in 50 mm)	Casting characteristics — Fluidity	Pressure tightness	Resistance to hot tearing	Corrosion resistance	Weldability	Machining
201.0		S*	T6	345	415	5.0	C	D	D	D	C	B
208.0		S	T533	105	185	1.5	B	B	B	D	B	C
213.0		PM	T533	185	220	0.5	B	C	C	E	C	B
	LM4	S and PM	T21	95	175	3.0	B	B	B	C	C	B
		PM	T6	230	295	2.0						
242.0	LM14	PM	T6	230	295	1.0	C	C	D	D	D	B
295.0		PM	T61	195	260	4.0	C	D	D	D	C	B
319.0	LM21	S	T21	125	185	1.0	B	B	B	C	B	C
		PM	T21	125	200	2.0						
332.0	LM13	S and PM	T61	240	260	0.5	B	C	A	C	B	B
		PM	T65	295	325	0.5						
355.0	LM16	S	T4	125	210	3.0	B	B	B	C	B	B
		PM	T4	140	245	6.0						
		PM	T6	235	280	1.0						
356.0	LM25	S	T6	205	230	4.0	B	A	A	B	B	C
		PM	T6	225	240	4.0						
360.0	LM9	S	T5	110	185	2.0	A	B	A	B	B	C
		S	T6	215	255	–						
		PM	T5	130	245	2.5						
		PM	T6	265	310	1.0						

Aluminum Association number	BS 1490 LM number	Casting process	Temper	0.2% proof stress (MPa)	Tensile strength (MPa)	Elongation (% in 50 mm)	Casting characteristics					
							Fluidity	Pressure tightness	Resistance to hot tearing	Corrosion resistance	Weldability	Machining
	LM6	S	F1	65	185	8.0	A	B	A	A	B	C
		PM	F1	90	205	9.0						
		D	F1	130	250	2.5						
413.0	LM20	D	F1	140	265	2.0	A	A	A	B	A	C
443.0	LM18	S	F1	65	130	5.0	B	A	A	A	B	C
		PM	F1	70	160	6.0						
514.0	LM5	S	F1	80	170	5.0	C	D	C	A	C	B
		PM	F1	80	230	10.0						
518.0		D	F1	130	260	10.0	D	E	C	A	A	A
520.0	LM10	S	T1	175	320	15.0	D	E	B	A	E	B
535.0		S	F	145	275	13.0	D	E	C	A	A	A
705.0		S	T1	130	240	9.0	D	C	E	B	D	A
707.0		S	T1	185	255	3.0	D	C	E	B	D	A
713.0		S	T5	175	235	4.0	D	C	E	B	D	A

Notes: S = sand casting; PM = permanent mould (gravity die) casting; D = pressure diecasting
* Results for sand cast alloys obtained from separately cast test bars
Ratings for casting characteristics A through to E in decreasing order of merit

(a)

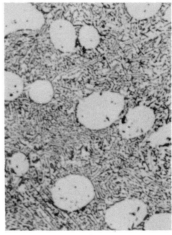

(b)

(c)

Fig. 4.1 As-cast Al-12Si alloy in the following conditions: a, unmodified; b, modified with sodium; c, excess phosphorus (courtesy Alcoa of Australia Ltd). ×400

the result that both ductility and tensile strength are much improved. The eutectic may also be refined by the process known as modification.

4.2.1 Modification

The widespread use of Al-Si alloys for other types of castings derives from the discovery by Pacz in 1920 that a refinement or modification of microstructure, similar to that achieved by rapid cooling, occurred when certain alkali fluorides were added to the melt prior to pouring (Fig. 4.1b).

Table 4.3   Mechanical properties of Al-13% Si alloys (from Thall, B. M. and Chalmers, B., *J. Inst. Met.*, **77**, 79, 1950)

Condition	Tensile strength (MPa)	Elongation (%)	Hardness (Rockwell B)
Normal sand cast	125	2	50
Modified sand cast	195	13	58
Normal chill cast	195	3.5	63
Modified chill cast	220	8	72

As shown in Table 4.3, mechanical properties may be substantially improved due to refinement of the microstructure and to a change to a planar interface during solidification which minimizes porosity in the casting. Modification of Al-Si alloys is now usually achieved by the addition of sodium salts or small quantities (0.005 to 0.015%) of metallic sodium to the melt although the actual amount of sodium needed may be as little as 0.001%. The mechanism by which the microstructure and, more particularly, the size and form of the silicon phase is modified has been the subject of much research. Controversy still remains although most theories involve possible effects of sodium on the nucleation and/or growth of eutectic silicon during solidification.

Sodium may depress the eutectic temperature by as much as $12°$ C and a finer microstructure is therefore to be expected because the rate of nucleation will be greater in the undercooled condition. Depression of the eutectic temperature implies that sodium reduces the potency of nucleating sites for the eutectic phases, notably silicon. It is known that silicon itself is readily nucleated at the surface of particles of the compound AlP which is formed by reaction of aluminium with impurity amounts of phosphorus. In fact, an excessive level of phosphorus can lead to the formation of a third, granular type of microstructure containing large particles of silicon which results in poor mechanical properties (Fig. 4.1c). Accordingly, a possible explanation for the behaviour of sodium is that it neutralizes the effect of phosphorus, probably by the preferential formation of the compound NaP. In a similar way, the phenomenon known as over-modification, whereby coarse silicon particles may reappear when an excess of sodium is present, has been attributed to formation of another compound, AlNaSi, which once again provides sites for the easy nucleation of silicon.

Strongest support for the alternative concept that sodium exerts its effect by restricting growth of silicon particles comes from the observation that these particles are interconnected within each cell, suggesting that there is no need for repeated nucleation to occur after each cell has formed. Many theories have been proposed to account for the possible effect of sodium on the growth of silicon. For example, it has been suggested that segregation of sodium at the interface of growing silicon particles immobilizes the growth steps at the surface of the crystals. Another theory postulates that silicon morphology is strongly growth-temperature dependent and the fact that sodium induces growth at a reduced temperature may cause the form of the silicon to be changed. Restricted growth theories alone do not, however, account for the coarsening of silicon in the presence of an excess of sodium and it seems probable that the effect of the modifying elements on the process of nucleation is the dominant factor.

The use of sodium presents founding problems such as reduced fluidity, but its major disadvantage arises through the rapid loss of the element by evaporation or oxidation. It is therefore necessary to add an excess amount and difficulties in controlling the content in the melt can lead to the under- or over-modification of the final castings. For the same reason, the effects of modification are lost if Al-Si castings are remelted which prevents foundries being supplied by pre-modified ingots. Research has, therefore, been directed at replacing sodium and considerable commercial success has been achieved by modifying with strontium, following work carried out in Australia and Germany. The addition of 0.03 to 0.05 % of this element as an Al-Sr or Al-Si-Sr master alloy also produces a refined Al-Si eutectic and the tensile properties of castings are comparable with those obtained when sodium is used. Loss of strontium during melting is much less and, moreover, the modified microstructure is retained after remelting. Over-modification is also less of a problem with strontium since excess amounts of this element are taken into the compound $SrAl_3Si_3$ which has comparatively little effect on the size of the silicon particles in the eutectic. Yet another advantage of strontium additions is that they supress formation of primary silicon in hypereutectic alloys which should improve their ductility and toughness. This effect is not observed when sodium is used.

4.2.2 Binary Al-Si alloys

Binary Al-Si alloys up to the eutectic composition retain good levels of ductility, providing the iron content is controlled to minimize formation of large, brittle plates of the compound α-AlFeSi. In this regard additions of manganese have been found to be beneficial. If the silicon content is below 8 %, modification is not necessary to achieve acceptable levels of ductility because the primary aluminium phase is present in relatively large amounts. The eutectic composition, which has a high degree of fluidity and low shrinkage on solidification, has particular application for thin-walled castings, e.g. Fig. 4.2. As a class, the alloys are used for sand and permanent mould castings for which strength is not a prime consideration, e.g. domestic cookware, pump casings and certain automobile castings, including water-cooled manifolds.

When as-cast alloys containing substantial amounts of silicon are subjected to elevated temperatures they suffer growth due to precipitation of silicon from solid solution. Dimensional stability can be achieved by heating for several hours in the temperature range 200 to 250° C prior to subsequent machining or use, and tempers of the T5 or T7 types should be given to castings which are to be used at temperatures of 150°C or above.

Fig. 4.2 Thin-walled Al-Si alloy casting

4.2.3 Higher strength alloys

Although binary Al-Si alloys show some response to heat treatment because the aluminium phase can be moderately supersaturated with silicon by rapid cooling, much greater strengthening is possible by the addition of other alloying elements such as copper and magnesium. Copper increases strength and improves machinability, although at the expense of reduced castability, ductility and corrosion resistance. Commercial Al-Si-Cu alloys have been available for many years and a compromise has been reached between these various properties (Table 4.2). Compositions lie mostly within the ranges 3 to 10.5 % silicon and 1.5 to 4.5 % copper. The higher silicon alloys (e.g. Al-10Si-2Cu) are used for pressure diecastings, whereas alloys with lower silicon and higher copper (e.g. Al-3Si-4Cu) are used for sand and permanent mould castings. The strength and machinability of some of these castings is often improved by artificial ageing (T5 temper). In general the Al-Si-Cu alloys are used for many of the applications listed for the binary alloys but where higher strength is needed. As with the wrought alloys, some compositions contain minor additions of elements such as bismuth and lead which improve machining characteristics.

More complex compositions are available where special properties are required. One example is the piston alloys for internal combustion engines, e.g. A332 (Al-12Si-1Cu-1Mg-2Ni) in which nickel, in particular, improves elevated temperature properties by forming stable intermetallic compounds that cause dispersion hardening. Another example is the range of hypereutectic compositions such as A390 (Al-17Si-4Cu-0.55Mg) which have been used for sand and permanent mould castings of all-aluminium alloy automotive cylinder blocks. Here the main direction of the developmental programmes has been the desire to eliminate using cast iron sleeves as cylinder liners which is the case in several production engines. In this regard it is necessary to incorporate, in the eutectic matrix, sufficient quantities of hard primary silicon particles to provide high wear resistance in the cylinders during service and yet keep the dispersion low enough so that serious machining problems are avoided. It is also desirable to ensure that the primary silicon is well refined. In this case, 0.01 to 0.03% of phosphorus is added which reacts with aluminium to form small, insoluble particles of AlP that then serve as nuclei on which silicon forms (Fig. 4.3).

Large quantities of sand and permanent mould castings are made from Al-Si-Mg alloys such as 356 or LM25 (Al-7Si-0.3Mg) in which the comparatively small additions of magnesium induce significant age hardening through precipitation of Mg_2Si in the aluminium matrix. For example, the yield strength of this alloy in the T6 condition is more than double that of the binary alloy containing a similar content of silicon. In addition they also display excellent corrosion resistance. The alloys find particular use for aircraft and automotive applications, one recent example being the lightweight wheels for sportscar enthusiasts. The critical nature of some of these applications has led to some recent studies of the relationships between microstructure and toughness, and values of fracture toughness as high as 30 MPa m$^{\frac{1}{2}}$ have been recorded for certain heat

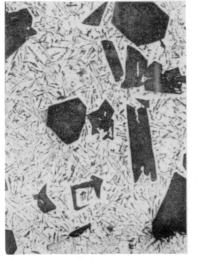
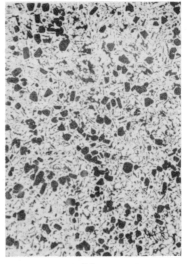

(a)   (b)

Fig. 4.3 Effect of minor additions of phosphorus in refining the size of the primary silicon plates in the hypereutectic alloy A390: a, no phosphorus; b, small additions of phosphorus (from Jorsted, J. L., *Trans. Met. Soc. AIME*, **242**, 1219, 1968). ×100

treated compositions which compare well with values expected for the higher strength, wrought alloys. This suggests that the possible replacement of some wrought components by these relatively cheaper castings may be a trend in the future. Examples of such a change are already available for some critical aircraft fittings such as the sand cast Al-Si-Mg engine pylon shown in Fig. 4.4.

4.3 Alloys based on the aluminium-copper system

Alloys with copper as the major alloying addition were the first to be widely used for aluminium castings although many have now been superseded. Most existing compositions contain additional alloying elements. As a group, these alloys may present casting problems, e.g. hot-tearing, and it is also essential to provide generous feeding during solidification to ensure soundness in the final product. The alloys respond well to age-hardening heat treatments.

Several compositions are available which have elevated temperature properties that are superior to all other classes of aluminium casting alloys. Examples are 238 (Al-10Cu-3Si-0.3Mg), which is used for permanent mould casting of the soleplates of domestic hand irons, and 242 (Al-4Cu-2Ni-1.5Mg), which has been used for many years for diesel engine pistons and air-cooled cylinder heads for aircraft engines. Each alloy relies on a combination of precipitation hardening together with dispersion hardening by intermetallic compounds to provide stability of strength and hardness at temperatures up to around 250°C.

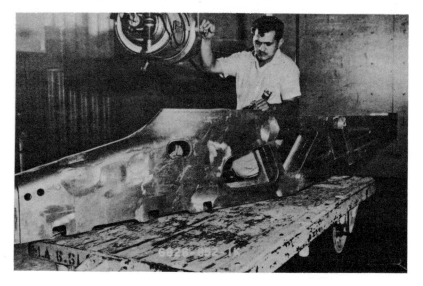

Fig. 4.4 Pylon for a fighter aircraft prepared by premium casting techniques (courtesy Defence Metals Information Center, Columbus, USA)

A more recent development has been the introduction of an Al-Cu-Ag-Mg alloy which has a particularly high response to age hardening. This alloy, which is known as KO-1 or 201 in the United States, has the nominal composition Al-4.7Cu-0.7Ag-0.3Mg and there is a similar alloy available in France called Avior which also contains 1.3 % zinc. The high response to ageing appears to arise because silver completely changes the precipitation process expected for the Al-Cu-Mg system, causing a monoclinic form of the phase θ(CuAl$_2$) to form as remarkably thin plates on the {111} matrix planes rather than the {100} planes (Fig. 4.5). Using premium quality casting techniques, guaranteed properties of 345 MPa proof stress and 415 MPa tensile strength with a minimum elongation of 5 % have been obtained from a variety of castings heat treated to the T6 temper, and values as high as 480 MPa proof stress and 550 MPa tensile strength with 10 % elongation have been recorded. These tensile properties are much higher than can be obtained with any other aluminium casting alloys and compare well with the high-strength wrought alloys. The alloys may be susceptible to stress-corrosion cracking in the T6 condition but resistance is greatly improved by heat treating to a T73 temper. Although the addition of silver is costly, the alloys may be cost-effective as a replacement for some wrought components.

4.4 Aluminium-magnesium alloys

This group contains several essentially binary alloys together with some more complex compositions based on Al-4Mg. Their special features are a high resistance to corrosion, good machinability and attractive appearance when anodized. Most show little or no response to heat treatment.

Casting characteristics are again less favourable than Al-Si alloys and more control must be exercised during melting and pouring because magnesium increases oxidation in the molten state. In this regard, special precautions are needed when sand casting as steam generated from moisture in the sand will react to produce MgO and hydrogen, which

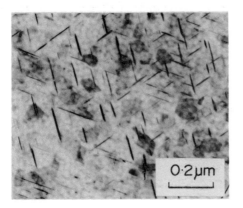

Fig. 4.5 Thin plates of a monoclinic form of the phase CuAl$_2$ on the {111} planes of the Al-Cu-Mg-Ag alloy 201, aged to peak hardness at 170°C (from Taylor, J. A. *et al.*, *Metal Science*, **12**, 478, 1978)

0·2 μm

results in roughening and blackening of the surface of the casting. This mould reaction can be reduced by adding about 1.5 % boric acid to the sand which forms a fused, glassy barrier towards steam produced within the mould. An alternative method is to add small amounts (0.03 %) of beryllium to the alloy, which results in the formation of an impervious oxide film at the surface. Beryllium also reduces general oxidation during melting and casting although special care must be taken because of the potentially toxic nature of BeO. As a group, Al-Mg alloys also require special care with gating, and large risers and greater chilling are needed to produce sound castings.

The magnesium contents of the binary alloys range from 4 to 10 %. Most are sand cast although compositions with 7 and 8 % magnesium have limited application for permanent mould and pressure diecastings. Al-10Mg responds to heat treatment and a desirable combination of high strength, ductility and impact resistance may be achieved in the T4 temper. The castings must be slowly quenched from the solution treatment temperature, otherwise residual stresses may lead to stress-corrosion cracking. In addition, the alloy tends to be unstable, particularly in tropical conditions, leading to precipitation of Mg_5Al_8 (perhaps Mg_2Al_3) in grain boundaries which both lowers ductility and may cause stress-corrosion cracking after a period of time. Consequently other alloys such as Al-Si-Mg tend to be preferred unless the higher strength and ductility of Al-10Mg are mandatory.

Casting characteristics are somewhat improved by ternary additions of zinc and silicon and alloys such as Al-4Mg-1.8Zn and Al-4Mg-1.8Si can be diecast for parts of simple design.

4.5 Aluminium-zinc-magnesium alloys

Several binary Al-Zn alloys have been used in the past but all are now obsolete except for compositions which are used as sacrificial anodes to protect steel structures in contact with sea water. Engineering alloys currently in use contain both zinc and magnesium, together with minor additions of one or more of the elements copper, chromium, iron, and manganese.

The as-cast alloys respond to ageing at room temperature and harden over a period of weeks. In this condition, or when artificially aged or stabilized after casting, the proof stress values range from 115 to 260 MPa and the tensile strengths range from 210 to 310 MPa, depending upon composition. Casting characteristics are relatively poor and the alloys are normally sand cast because the use of permanent moulds tends to cause hot-cracking.

One advantage of the alloys is that eutectic melting points are relatively high which makes them suitable for castings that are to be assembled by brazing. Other characteristics are good machinability, dimensional stability and resistance to corrosion. The alloys are not recommended for use at elevated temperatures because over-ageing causes rapid softening.

4.6 Joining

Most aluminium casting alloys can be arc welded in a protective atmosphere of an inert gas, e.g. argon, provided they are given the correct edge preparation. Ratings of weldability were included in Table 4.2. In addition some surface defects and service failures in sand and permanent mould castings may be repaired by welding. Filler metals are selected which are appropriate to the compositions of alloy castings with 4043 (Al-5Si) and 5356 (Al-5Mg-0.1Mn-0.1Cr) being commonly used. With respect to joining of castings by brazing, similar conditions apply to those discussed for wrought aluminium alloys in Section 3.6.2.

Further reading

Van Horn, K. (ed), *Aluminum,* Volumes 1–3, American Society for Metals, Cleveland, 1967

Casting: Kaiser Aluminum, Second edition, Kaiser Center, Oakland, 1965

The Properties of Aluminium and its Alloys, Aluminium Federation, Birmingham, 1973

Varley, P. C., *The Technology of Aluminium and its Alloys*, Newnes-Butterworths, London, 1970

Metals Handbook, Volume 5, Part B, Melting and Casting, American Society for Metals, 1972; p. 389

Fortina, G., Guidelines for the use of primary aluminium castings, *Aluminio E. Nuova Metallurgia,* **44**, 234, 1979

Grube, K. R. *et al.*, Premium quality aluminum castings, *Defence Metals Information Center Report 211*, Battelle Memorial Institute, Columbus, 1965

Smith, R. W., *Solidification of Metals*, Iron and Steel Institute Publication P110, 1968; p. 224

Hogan, L. M. and Jenkinson, D. C., The modification of aluminium-silicon alloys with strontium, *J. Crystal Growth*, **28**, 171, 1975

Alker, K. and Hielscher, V., Experiences with the permanent modification of an aluminium-silicon casting alloy, *Aluminium*, **48**, 362, 1972

Jorsted, J. L., Development of the hypereutectic Al-Si die casting alloy used in the Vega engine block, *Trans. Met. Soc. AIME*, **242**, 1219, 1968

5
Magnesium alloys

5.1 Introduction to alloying behaviour

Magnesium is readily available commercially with purities exceeding 99.8 % but is rarely used for engineering applications in its unalloyed form. It has an hexagonal lattice structure (Fig. 5.1) and its alloying behaviour is notable for the variety of elements with which it will form solid solutions. In this regard aluminium, zinc, lithium, cerium, silver, zirconium and thorium are examples of metals that are present in commercial alloys. Binary alloy systems with metals of commercial significance fall into two categories, the alloying additions being listed in order of decreasing atomic solid solubility with weight percentages being shown in brackets:

Peritectic systems: Mn (3.4 %), Zr (3.8 %)
Eutectic systems: Li (5.7 %), Al (12.7 %), Ag (15.5 %), Zn (8.4 %), Nd ($\sim$ 3 %), Th (4.5 %), Ce (0.85 %).

The earliest commercial alloying elements were aluminium, zinc, and manganese and Mg-Al-Zn castings were used extensively in Germany during the First World War. These alloys suffered from corrosion problems in wet or moist environments which were much reduced with the discovery in 1925 that small additions (0.2 %) of manganese increased corrosion resistance. The role of this addition was later found to involve the removal of iron and certain other impurities into relatively harmless intermetallic compounds. Alloys based on the Mg-Al-Zn system have remained the principal materials for magnesium casting alloys for use at ambient

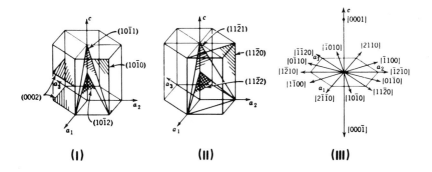

(I)   (II)   (III)

Fig. 5.1 Principal planes and directions in the magnesium unit cell

127

Table 5.1 Nominal composition, typical tensile properties and characteristics of selected magnesium casting alloys

ASTM desig-nation	British desig-nation	Nominal composition								Condition	Tensile properties			Characteristics
		Al	Zn	Mn	Zr	RE (MM)	RE (Nd)	Th	Ag		0.2% proof stress (MPa)	Tensile strength (MPa)	Elongation (%)	
AZ63		6	3	0.3						As-sand cast	75	180	4	Good room temperature strength and ductility
										T6	110	230	3	
AZ81	A8	8	0.5	0.3						As-sand cast	80	140	3	Tough, leak-tight castings. With 0.0015 Be used for pressure diecasting
										T4	80	220	7	
AZ91	AZ91	9.5	0.5	0.3						As-sand cast	95	135	2	General purpose alloy used for sand and diecastings
										T4	80	230	4	
										T6	120	200	3	
										As-chill cast T4	100	170	2	
										T6	80	215	5	
										T6	120	215	2	
ZK51	Z5Z		4.5		0.7					T5	140	235	5	Sand castings, good room temperature strength and ductility
ZK61			6		0.7					T5	175	275	5	As for ZK51
ZE41	RZ5		4.2		0.7	1.3				T5	135	180	2	Sand castings, good room temperature strength, improved castability

ASTM desig-nation	British desig-nation	Nominal composition								Condition	Tensile properties			Characteristics
		Al	Zn	Mn	Zr	RE (MM)	RE (Nd)	Th	Ag		0.2% proof stress (MPa)	Tensile strength (MPa)	Elongation (%)	
EZ33	ZRE1		2.7		0.7	3.2				Sand cast T5 Chill cast T5	95 100	140 155	3 3	Good castability, pressure tight, weldable, creep resistant to 250°C
HK31	MTZ				0.7			3.2		Sand cast T6	90	185	4	Sand castings, good castability, weldable, creep resistant to 350°C
HZ32	ZT1		2.2		0.7			3.2		Sand or chill cast } T5	90	185	4	As for HK31
OE22	MSR				0.7		2.5		2.5	Sand or chill cast } T6	185	240	2	Pressure tight and weldable, high proof stress to 250°C
QH21	QH21				0.7		1	1	2.5	As-sand cast T6	185	240	2	Pressure tight, weldable, good creep resistance and proof stress to 300°C

● Values quoted for tensile properties are for separately cast test bars and may not be realized in certain parts of castings

RE = rare earth element; MM = misch metal; Nd = neodymium

Table 5.2 Nominal composition, typical tensile properties and characteristics of selected wrought magnesium alloys

ASTM desig- nation	British desig- nation	Nominal composition						Condition	Tensile properties			Characteristics
		Al	Zn	Mn	Zr	Th	Li		0.2% proof stress (MPa)	Tensile strength (MPa)	Elongation (%)	
M1	AM503			1.5				Sheet, plate F Extrusions F Forgings F	70 130 105	200 230 200	4 4 4	Low- to medium-strength alloy, weldable, corrosion resistant
AZ31	AZ31	3	1	0.3 (0.20 min)				Sheet, plate O H24 Extrusions F Forgings F	120 160 130 105	240 250 230 200	11 6 4 4	Medium-strength alloy, weldable, good formability
AZ61	AZM	6.5	1	0.3 (0.15 min)				Extrusions F Forgings F	180 160	260 275	7 7	High-strength alloy, weldable
AZ80	AZ80	8.5	0.5	0.2 (0.12 min)				Forgings T6	200	290	6	High-strength alloy
ZM21	ZM21		2	1				Sheet, plate O H24 Extrusions Forgings	120 165 155 125	240 250 235 200	11 6 8 9	Medium-strength alloy, good form- ability, good damping capacity
LA141		1.2		0.15 min			14	Sheet, plate T7	95	115	10	Ultra-light weight (*d* 1.35)

ASTM desig-nation	British desig-nation	Nominal composition						Condition	Tensile properties			Characteristics
		Al	Zn	Mn	Zr	Th	Li		0.2% proof stress (MPa)	Tensile strength (MPa)	Elongation (%)	
ZK31	ZW3		3		0.6			Extrusions T5	210	295	8	High-strength alloy some weldability
								Forgings T5	205	290	7	
ZK61			6		0.8			Extrusions F	210	285	6	High-strength alloy
								T5	240	305	4	
								Forgings T5	?160	275	7	
HK31					0.7	3.2		Sheet, plate H24	170	230	4	High creep resistance to 350°C, weldable
								Extrusions T5	180	255	4	
HM21				0.8		2		Sheet, plate T8	135	215	6	High creep resistance to 350°C, short time exposure to 425°C, weldable
								T81	180	255	4	
								Forgings T5	175	225	3	
HZ11	ZTY		0.6		0.6	0.8		Extrusions F	120	215	7	Creep resistance to 350°C, weldable
								Forgings	130	230	6	

temperatures. The earliest wrought alloy was Mg-1.5Mn which was used for sheet, extrusions and forgings, but this material has now been superseded.

In the period between the two World Wars, difficulties were encountered with magnesium alloy castings because they tended to have large and variable grain size that often resulted in poor mechanical properties, microporosity and led to excessive directionality of properties in wrought components. In particular, values of proof stress tended to be consistently low relative to tensile strength. In 1937, Sauerwald of IG Farbenindustrie in Germany discovered that zirconium had an intense grain refining effect on magnesium, but another ten years elapsed before a reliable method was developed whereby this metal could be alloyed with magnesium. Paradoxically, zirconium could not be used in the existing commercial alloys because it was removed from solution due to the formation of stable compounds with both aluminium and manganese. This led to the evolution of a complete new series of cast and wrought, zirconium-containing alloys having much improved mechanical properties at both room and elevated temperatures (Tables 5.1 and 5.2). These alloys are now widely used in the aerospace industries.

Cast magnesium alloys have predominated over wrought products, particularly in Europe, where they have comprised 85–90 % of magnesium alloys produced in recent years. Forgings and extrusions are the most widely used wrought products. In addition, there is a number of applications for plate in the United States where higher capacity rolling mills are available and a greater market exists in the aerospace industries.

Several magnesium alloys are amenable to age hardening although their response is significantly less than is observed in some aluminium alloys. The precipitation processes are frequently complex and are not completely understood. Probable precipitation sequences in alloys of commercial interest are shown in Table 5.3 (see p. 142). A feature of the ageing process in most alloys is that one stage involves the formation of an ordered, hexagonal precipitate with a DO_{19} (Mg_3Cd) crystal structure that is coherent with the magnesium lattice. This structure is analogous to the well known θ'' (GP zones 2) phase that may form in aged Al-Cu alloys and it is commonly found in alloys in which there is a large difference in the atomic sizes of the constituents. The phase contributes to hardening in those magnesium alloys in which it forms and it is present at peak hardness over a wide temperature range.

The DO_{19} cell has an a axis twice the length of the a axis of the magnesium matrix whereas the c axes are the same. The precipitate forms as plates or discs parallel to the $<0001>_{Mg}$ directions which lie along the $\{10\bar{1}0\}_{Mg}$ and $\{11\bar{2}0\}_{Mg}$ planes. In this regard it is significant to note that alternate $(10\bar{1}0)$ and $(11\bar{2}0)$ planes in a structure of composition Mg_3X consist of all magnesium atoms (Fig. 5.2). Thus, the formation of a low energy interface along these planes is to be expected since only second nearest neighbour bonds need to be altered. This structural feature would account for the fact that the phase is relatively stable over a wide

Fig. 5.2   Model showing low energy interfaces along the $\{10\bar{1}0\}$ and $\{11\bar{2}0\}$ planes of precipitates having the DO_{19} structure (Mg_3X). White balls = magnesium atoms, black balls = solute atoms (from Gradwell, K. J., *Precipitation in a High Strength Magnesium Casting Alloy*, PhD Thesis, University of Manchester, 1972)

temperature range and it may be the most significant factor in promoting creep resistance in those magnesium alloys in which it occurs.

5.2 Melting and casting

5.2.1 Melting

It is usual for magnesium to be melted in mild steel crucibles for both the alloying and refining or cleaning stages before producing cast or wrought components. Unlike aluminium and its alloys, the presence of an oxide film on molten magnesium does not protect the metal from further oxidation. On the contrary, it accelerates this process. Melting is complete at or below 650°C and the rate of oxidation of the molten metal surface increases rapidly with rise in temperature such that above 850°C a freshly exposed surface spontaneously bursts into flame. Consequently, suitable fluxes or inert atmospheres must be used when handling molten magnesium and its alloys.

The reactivity of magnesium limits the choice of fluxes to chlorides and fluorides of the alkali and alkaline earth metals, including magnesium, and to certain inert oxides. These fluxes fall into two classes: thinly fluid and thickened or 'inspissated' fluxes. The first class is used during the melting stage and comprises a mixture of chlorides such as $MgCl_2$ with KCl or NaCl. In Britain, it is usual practice then to replace this flux by an inspissated flux which contains, in addition, a mixture of CaF_2, MgF_2 and MgO. This flux forms a coherent, viscous cake that excludes air during subsequent alloying and refining, and which can be readily drawn aside when pouring. Prior to pouring, the melt is stirred to promote flux removal of the oxides and chlorides that are present as a suspension, the removal of chlorides being essential because of their adverse effect on the corrosion resistance of magnesium and its alloys. The fact that several of the chlorides including $MgCl_2$ are deliquescent further increases their corrosive effect when the alloys are exposed to the atmosphere. During pouring, it is common to dust the molten metal with sulphur to minimize oxidation.

Most alloying elements are now added in the form of master alloys or hardeners. Zirconium has presented special problems as early attempts to use either zirconium metal or a Mg-Zr hardener were ineffective. Success was achieved eventually by means of mixtures of reducible zirconium halides, e.g. ones containing fluorozirconate, K_2ZrF_6, together with large amounts of $BaCl_2$ to increase the density of the salt reaction products. These salt mixtures were supplied under licence to foundries. Subsequently it was found possible to prepare hardener alloys from the weighted salt mixtures and proprietary hardeners made in this way are now used for adding zirconium to magnesium alloy melts. Prior to the use of $BaCl_2$, severe problems were encountered with persistent flux inclusions which arose through entrainment of salt reaction products in the melt and could not be removed by pre-solidification or any flux-refining step.

As with aluminium, hydrogen is the only gas that dissolves in molten magnesium although it is less of a problem in this case because of its comparatively high solid solubility (average of about 30 ml $100 g^{-1}$). The main source of hydrogen is from water vapour in damp fluxes or corroded scrap/ingot, so pick-up can be minimized by taking adequate precautions with these materials. A low hydrogen content reduces the tendency to gas porosity which is common in Mg-Al and Mg-Al-Zn alloys and these materials should be degassed with chlorine. The optimum temperature for degassing is 725 to 750°C. If the melt is below 713°C then solid $MgCl_2$ will form which gives little protection from burning, whilst at temperatures much above 750°C, magnesium losses through reaction with chlorine become excessive. Gas porosity is not normally a problem with zirconium-containing alloys since zirconium will itself remove hydrogen as ZrH_2 and it is generally unnecessary to degas these alloys. However, it should be noted that such a treatment does improve the tensile properties of certain Mg-Zn-Zr alloys, presumably by minimizing the loss of zirconium as the insoluble ZrH_2. In such a case the degassing operation is completed before zirconium is added.

5.2.2 Grain refinement

Although grain refinement is carried out during melting, its importance and complexity in magnesium alloys merits separate consideration. Quite different practices are needed depending upon the presence or absence of zirconium.

The group of alloys based mainly on the Mg-Al system tends to have large and variable grain size. The first method devised to control grain size was to superheat the melt to a temperature of 850°C and above for periods of about 30 min, after which the melt was quickly cooled to the normal casting temperature and poured. A comparatively fine grain size was achieved with fair success, although the reasons for the effect are somewhat obscure. The probable explanation is that foreign nuclei with suitable crystal structures such as Al_4C_3 precipitate on cooling to the casting temperature and act as nuclei for the magnesium grains during subsequent solidification. The superheating effect is only significant in Mg-Al alloys and presents problems because crucible and furnace lives are reduced and power requirements are increased.

An alternative technique was developed in Germany in which a small quantity of anhydrous $FeCl_3$ was added to the melt (Elfinal process) and grain refinement was attributed to nucleation by iron-containing compounds. This method also had its disadvantages because the deliquescent nature of $FeCl_3$ made it hazardous, and the presence of as little as 0.005 % iron could decrease the corrosion resistance of the alloys. The addition of manganese was made to counter this latter problem but effectively prevented grain refinement by $FeCl_3$.

The method in current use for alloys containing aluminium as a major alloying element is to add volatile carbon-containing compounds to the melt and hexachlorethane (0.025 to 0.1 % by weight) is commonly used in the form of small briquettes which are held at the bottom of the melt whilst they dissociate into carbon and chlorine. Grain refinement is attributed to inoculation of the melt with Al_4C_3 or $AlN.Al_4C_3$. Release of chlorine causes some degassing of the melt which is a further advantage of the method.

The way that zirconium grain refines magnesium is uncertain although it may be noted that the lattice parameters of hexagonal α-zirconium ($a = 0.323$ nm, $c = 0.514$ nm) are very close to those of magnesium ($a = 0.320$ nm, $c = 0.520$ nm). This has led to the suggestion that zirconium may nucleate magnesium and microprobe analysis has revealed the presence of zirconium-rich cores in the centres of magnesium grains. It has been proposed that a peritectic mechanism is involved in which zirconium particles, separating from the liquid, react with it at the peritectic temperature, thereby acquiring a layer of zirconium-enriched solid solution and serving as nuclei during solidification. However, as shown in Fig. 5.3, such a reaction would not be expected unless the zirconium content exceeded 0.58 %. Since significant grain refinement occurs with lower amounts of this element, the possibility of nucleation by zirconium *per se* or a zirconium compound cannot be excluded.

Many factors operate which can precipitate zirconium from molten magnesium and it is only possible to maintain a maximum content in solution if a considerable excess of zirconium is present at the bottom of the melt. On transfer to another crucible, the soluble zirconium content of the melt will at once begin to fall, so all zirconium-containing castings should be poured directly from the crucible in which the alloying is done. If transfer is necessary, then the alloy will need to be replenished with zirconium if castings with a guaranteed fine grain size are to be obtained.

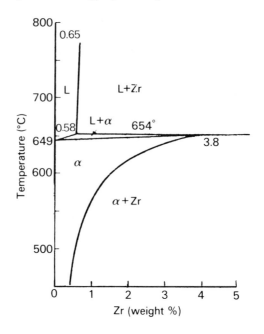

Fig. 5.3   Section of the Mg-Zr phase diagram

5.2.3 Casting

Sand castings constitute the largest single use of magnesium alloys and some general principles must be followed which are dictated by the physical properties and chemical reactivity of magnesium.

(i) Suitable inhibitors must be added to the moulding sand in order to avoid reaction between molten magnesium and moisture which would liberate hydrogen. For green sand or sand gassed with carbon dioxide to aid bonding, sulphur is used, whereas for synthetic sands compounds such as KBF_4 and $KSiF_6$ are also added. Boric acid is also used for some sands, both as a moulding aid and as a possible inhibitor through its tendency to coat the sand grains.

(ii) Metal flow should be as smooth as possible to minimize oxidation.

(iii) Because of the low density of magnesium, there is a relatively small pressure head in risers and sprues to assist the filling of moulds. Sands need to be permeable and the mould must be well-vented to allow the expulsion of air.

(iv) Magnesium has a relatively low volume heat capacity which necessitates provision of generous risers to maintain a reservoir of hotter metal. Because of this need to make special provision by way of risers and other feeding devices, the volume ratio of poured metal to actual castings may average as much as 4:1 for magnesium alloys.

The permanent mould casting process has features in common with sand casting although the faster rates of solidification promote finer grain size and uniformly higher mechanical properties. Pressure diecastings are produced by both hot- and cold-chamber methods. The low volume heat capacity of magnesium proves to be an advantage in this as less heat needs to be extracted by the metal mould than is the case with zinc or aluminium, thereby offering an opportunity for increased productivity. Relative speeds for zinc, aluminium, and magnesium cold-chamber pressure diecasting have been quoted as 1.0:1.6:1.9 respectively.

Ingots for producing wrought products were formerly produced in permanent metal moulds, but semi-continuous direct-chill methods are now in general use. These methods are broadly similar to those described for aluminium alloys and illustrated in Fig. 3.1.

5.3 Alloy designations and tempers

No international code for designating magnesium alloys exists although there has been a trend towards adopting the method used by the American Society for Testing Materials. In this system, the first two letters indicate the principal alloying elements according to the following code: A—aluminium; B—bismuth; C—copper; D—cadmium; E—rare earths; F—iron; G—magnesium; H—thorium; K—zirconium; L—lithium; M—manganese; N—nickel; P—lead; Q—silver; R—chromium; S—silicon; T—tin; Y—antimony; Z—zinc. The letter corresponding to the element present in greater quantity in the alloy is used first, and if they are equal in quantity the letters are listed alphabetically. The two (or one) letters are followed by numbers which represent the nominal compositions of these principal alloying elements in weight %, rounded off to the nearest whole number, e.g. AZ91 indicates the alloy Mg-9Al-1Zn the actual composition ranges being 8.3–9.7 Al and 0.4–1.0 Zn. A limitation is that information concerning other intentionally added elements is not given, and the system may need to be modified on this account. Suffix letters A, B, C etc. refer to variations in composition within the specified range and X indicates that the alloy is experimental. This system will be used when discussing alloys in this book, although the designations of equivalent British alloys are listed in the tables.

The heat treated or work-hardened conditions, i.e. tempers, of alloys are specified in the same way as has been described for aluminium alloys in Section 3.2.

5.4 Zirconium-free casting alloys

5.4.1 Magnesium-aluminium alloys

The Mg-Al system has been the basis of the most widely used magnesium casting alloys since these materials were introduced in Germany during the First World War. Most alloys contain 8 to 9% aluminium with small amounts of zinc, which gives some increase in tensile properties, and

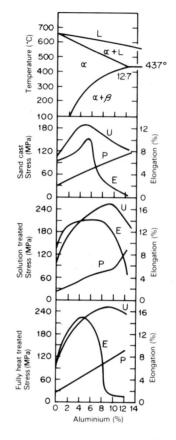

Fig. 5.4 Relation of properties to constitution in the Mg-Al alloys, showing effects of composition and heat treatment on the tensile properties of sand cast A8 alloy. P = 0.2% proof stress, U = tensile strength, E = elongation (from Fox, F. A., *J. Inst. Metals*, **71**, 415, 1945)

manganese, e.g. 0.3%, which improves corrosion resistance (Section 5.1). The presence of aluminium requires that the alloys be grain refined by superheating or inoculation. A section of the Mg-Al phase diagram together with a summary of the effects of composition and heat treatment on tensile properties of an alloy similar to AZ80A is shown in Fig. 5.4 and other property data are included in Table 5.1.

In the as-cast state, the β-phase $Mg_{17}Al_{12}$, sometimes referred to as Mg_4Al_3, appears in alloys containing more than 2% aluminium. A network of the β-phase forms around the grain boundaries as the aluminium content is increased and the ductility decreases rapidly above 8% (Fig. 5.5a). In more slowly-cooled castings, discontinuous precipitation also occurs at grain boundaries with formation of a cellular or pearlitic structure (Fig. 5.5b). Annealing at around 420°C causes the cellular constituent and some

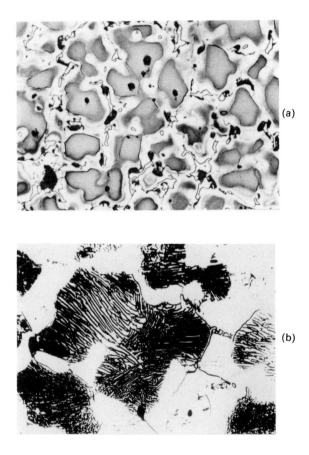

(a)

(b)

Fig. 5.5 Cast structures in the alloy AZ80: a, chill cast alloy with the β-phase ($Mg_{17}Al_{12}$) present in the grain boundaries. Note also the interdendritic aluminium-rich coring (white) around the edges of the α-grains (grey). ×200; b, discontinuous precipitation in more slowly-cooled alloys. ×500 (courtesy Magnesium Elektron Ltd)

of the β-phase to redissolve, leading to solid solution strengthening, and both tensile strength and ductility are significantly improved. Subsequent ageing at 150 to 250° C results in some precipitation of β throughout the grains and a further small increase in tensile strength.

Cast Mg-Al alloys show some susceptibility to microporosity but otherwise have good casting qualities and a satisfactory resistance to corrosion. They are suitable for use at temperatures up to 110–120°C. For higher temperatures up to 150 to 170°C, alloys with lower aluminium contents (e.g. 4%) and containing 0.8% silicon show higher creep strength, although fluidity is reduced. The lower aluminium content also reduces corrosion resistance but this effect can be countered by the addition of 0.2 % antimony. Although binary Mg-Al alloys have been used in the past for a wide range of castings, they now find comparatively little commercial application.

5.4.2 Magnesium-aluminium-zinc alloys

The addition of zinc to Mg-Al alloys causes some strengthening, although the amount added has been limited because susceptibility to hot-cracking during solidification is increased (Fig. 5.6). This applies particularly to diecastings. The level of zinc is inversely related to the aluminium content and two common alloys are AZ63 and AZ91, the latter alloy having a relatively high fatigue strength. These alloys are more widely used than the binary Mg-Al alloys. A well-known example is the crankcase for the Volkswagen motor car where substitution of the magnesium alloy AZ81 for cast iron provided a weight saving of 50 kg, which was critical in the design of a vehicle with a rear engine.

Recently it has been shown that compositions containing relatively high zinc contents have attractive diecasting characteristics, and an alloy designated AZ88 (Mg-8Al-8Zn) is claimed to have sufficient fluidity to be used for pressure diecastings. Fluidity and corrosion resistance are claimed to exceed those of the traditional alloys, e.g. AZ91. The new range of potential castable alloys is shown in Fig. 5.6 and compositions with zinc contents as high as 12 %, e.g. Mg-12Zn-4Al, are now being investigated for diecastings because the lower aluminium content improves subsequent plating characteristics.

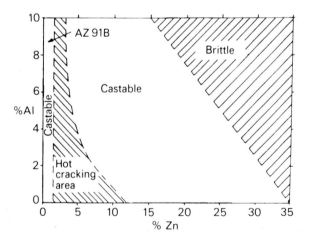

Fig. 5.6 Potential composition ranges of Mg-Al-Zn alloys for die-casting (from Foerster, G. S., *Proc. 33rd Annual Meeting Int. Magnesium Association*, Montreal, 1976)

5.4.3 Magnesium-zinc alloys

Magnesium-zinc alloys respond to age hardening but they are not amenable to grain refining by superheating or inoculation, and are susceptible to microporosity. Consequently they are not used for commercial castings.

As shown in Table 5.3, the ageing process is complex and may involve four stages. The GP zones solvus for the alloy Mg-5.5Zn lies between 70 and 80°C, and pre-ageing below this solvus, before ageing at a higher temperature (e.g. 150°C), refines the size and dispersion rods of the coherent $MgZn_2$ phase which then form from the GP zones. Maximum hardening is associated with the presence of this coherent phase.

5.5 Zirconium-containing casting alloys

The maximum solubility of zirconium in molten magnesium is 0.6 % and as binary Mg-Zr alloys are not sufficiently strong for commercial applications, the addition of other alloying elements has been necessary. The selection of these elements has been governed by three main factors:
(i) compatability with zirconium
(ii) founding characteristics
(iii) properties desired of the alloy.
With regard to (iii), improved tensile properties including higher ratios of proof stress to tensile strength, and increased creep resistance have been the two principal objectives. These have been decided primarily by requirements of the aerospace industries. Developments in various countries have followed similar lines and Table 5.1 and Figs. 5.7 and 5.8 summarize the mechanical properties of the more common alloys.

5.5.1 Magnesium-zinc-zirconium alloys

The ability to grain refine Mg-Zn alloys with zirconium led to the introduction of alloys, e.g. ZK51 (Mg-4.5Zn-0.7Zr) and the higher strength ZK61 (Mg-6Zn-0.7Zr), which are normally used in the T5 and T6 tempers respectively. However, the fact that these alloys are susceptible to microporosity and are not weldable has severely limited their practical application.

5.5.2 Magnesium-rare earth-zinc-zirconium alloys

The magnesium-rich regions of the phase diagrams for the major magnesium-rare earth (RE) systems, e.g. Mg-Ce, Mg-Nd and Mg-La, are all similar. Each is a simple eutectic and the same applies when naturally occurring cerium misch metal (average composition 49 %Ce, 26 %La, 19 %Nd, 6 %Pr) is added to magnesium. The alloys have good casting properties because the presence of the relatively low melting point eutectics as networks in the grain boundaries tends to suppress microporosity. The strength of the alloys may be raised to commercially acceptable levels by

Table 5.3 Probable precipitation processes in magnesium alloys

Alloy system	Precipitation process			
Mg-Al	SSSS → $Mg_{17}Al_{12}$ equilibrium precipitate nucleated on $(0001)_{Mg}$ (incoherent)			
Mg-Zn	SSSS → GP zones Discs $//\{0001\}_{Mg}$ (coherent)	→ $MgZn_2$ Rods $\perp\{0001\}_{Mg}$ cph $a = 0.52$ nm $c = 0.85$ nm (coherent)	→ $MgZn_2'$ Discs $//\{0001\}_{Mg}$ $(11\bar{2}0)_{MgZn_2}//(10\bar{1}0)_{Mg}$ cph $a = 0.52$ nm $c = 0.848$ nm (semi-coherent)	→ Mg_2Zn_3 Trigonal $a = 1.724$ nm $b = 1.445$ nm $c = 0.52$ nm $\gamma = 138°$ (incoherent)

Mg-RE(Nd)	SSSS → GP zones
	(Mg-Nd)
	Plates $//\{10\bar{1}0\}_{Mg}$
	(coherent)
		→ β''
		Mg_3Nd?
		cph
		DO_{19} superlattice
		Plates
		$(0001)_{\beta''}//(0001)_{Mg}$
		$\{10\bar{1}0\}_{\beta''}//\{10\bar{1}0\}_{Mg}$
			→ β'
			Mg_3Nd
			fcc
			Plates
			$a = 0.736$ nm
			$(011)_{\beta'}//(0001)_{Mg}$
			$(\bar{1}\bar{1}\bar{1})_{\beta'}//\{\bar{2}110\}_{Mg}$
			(semi-coherent)
				→ β
				$Mg_{12}Nd$
				bct
				$a = 1.03$ nm
				$c = 0.593$ nm
				(incoherent)

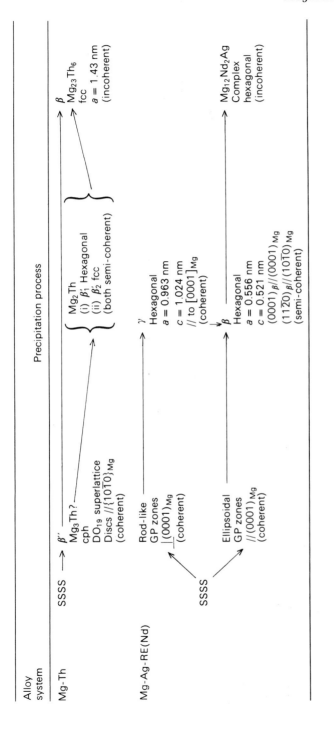

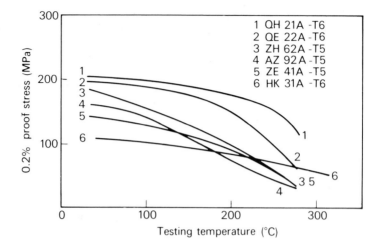

Fig. 5.7   Effect of temperature on the 0.2% proof stress of sand cast magnesium alloys (from Unsworth, W. H., *Proc. Conference International Magnesium Association*, 1977)

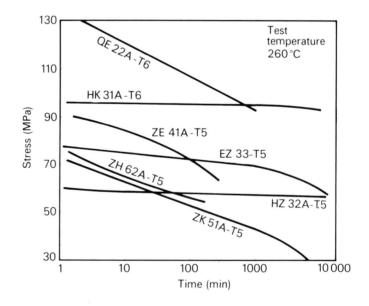

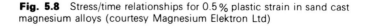

Fig. 5.8   Stress/time relationships for 0.5% plastic strain in sand cast magnesium alloys (courtesy Magnesium Elektron Ltd)

adding 0.7% zirconium to refine grain size but Mg-RE-Zr alloys have now been largely superseded by the stronger quarternary Mg-RE-Zn-Zr alloys, e.g. EZ33 which has the nominal composition Mg-3RE-2.5Zn-0.6Zr.

As mentioned earlier, Mg-Zn alloys respond to age hardening but have relatively poor casting qualities and are not weldable. These problems are largely overcome when RE elements are added, although the tensile properties that may be developed at room temperature in the binary alloys are reduced because some of the zinc is removed from solution through the formation of stable Mg-Zn-RE phases in grain boundaries. However, Mg-RE-Zn-Zr alloys such as EZ33 and ZE41 show good creep resistance up to 250°C and are widely used for castings exposed to elevated temperatures, e.g. Fig. 5.9.

In the as-cast condition, the alloys have cored α-grains which are surrounded by a grain boundary network. The eutectic tends to be divorced, so this network can comprise relatively massive amounts of the brittle

Fig. 5.9 Cutaway section of the magnesium alloy gearcase in the Sud-Aviation Super Frelon helicopter. Casting is made from the Mg-Zn-RE-Zr alloy ZE41 and weighs 133 kg as-cast and 111 kg finish machined (courtesy Magnesium Industry Council)

Fig. 5.10   Mg-RE-Zn-Zr alloy EZ33 as-cast and aged 8 h at 350° C (T5 temper). ×500

second phase (Fig. 5.10). An alloy such as EZ33 is normally used in the T5 condition and ageing causes precipitation to occur within the grains (Fig. 5.10). The good creep resistance of the alloys containing RE elements is attributed primarily to the strengthening effect of the precipitate in the grains, with the grain boundary phases perhaps contributing by reducing grain boundary sliding.

The precipitation processes in Mg-RE systems are not completely understood. Particular attention has been paid to Mg-Nd alloys in which four stages have been detected (Table 5.3). Most hardening is associated with formation of the coherent β''-phase which has the ordered DO_{19} structure and is probably Mg_3Nd. Loss of coherency of this phase occurs close to 250°C and is associated with a marked increase in creep rate. The β'-phase is thought to have a hexagonal structure and is nucleated on dislocation lines when the alloys are aged in the temperature range 200–300°C. It is possible that the MgRE (Ce) alloys have a similar ageing sequence which leads to the eventual formation of an equilibrium precipitate that is either $Mg_{12}Ce$ or $Mg_{27}Ce_2$.

The role of zinc with respect to strengthening is uncertain. It is likely that independent formation of Mg-Zn precipitates occurs although part of the zinc is associated with RE elements in constituents that form in grain boundaries. The effect of zirconium is thought to be confined to its role in grain refinement.

It might be thought that by starting with a high zinc content, e.g. 6%, and adding sufficient rare earth metals, e.g. 3%, an alloy could be produced which could combine the high tensile properties of ZK51 with the castability and freedom from microporosity of EZ33. Unfortunately such an alloy composition proves to be unusable since the low solidus temperature precludes solution heat treatment and the elongation, at 3% RE, is almost nil. Realizing that the low ductility was connected with a brittle grain boundary phase, it occurred to P. A. Fisher to decompose the latter by diffusing hydrogen into the alloy at a high temperature, thereby precipitating the RE metals as hydrides and enabling the zinc to be taken into solution (Fig. 5.11a, b). After final precipitation treatment in which a needle-like phase forms in the grains (Fig. 5.11c), castings in this alloy show high tensile properties with freedom from microporosity, together with high elongation values and outstanding fatigue resistance, which is an altogether remarkable combination of properties. The effect of the hydrogen treatment on tensile properties is shown for the Mg-6Zn-2RE composition in Table 5.4.

The final alloy developed in this way (ZE63) has the composition Mg-5.8Zn-2.5RE-0.7Zr. It has found limited but important usage in the aircraft industry. The rate of penetration of hydrogen is about 6 mm in 24 h at a temperature of 480° C and a pressure of 1 atm. Penetration can be accelerated by increasing the gas pressure, but the slowness of the hydriding step has hitherto restricted use of the alloy to castings with fairly thin sections.

Individual RE elements have differing effects on the response of magnesium alloys to heat treatment with the mechanical properties, generally increasing in the order of lanthanum, misch metal, cerium and didymium (72% neodymium). This effect is demonstrated for creep extension curves (Fig. 5.12) and probably corresponds to an increasing solid solubility of the elements in magnesium. It should also be noted that cerium is in fact marginally better than misch metal over the useful addition range of these elements, i.e. below 3%, but the expense of separating individual RE elements from misch metal is normally not justified.

5.5.3 Alloys based on the magnesium-thorium system

The addition of thorium also confers increased creep resistance to magnesium alloys, and cast and wrought materials which can be used in service at temperatures up to 350° C have been commercially available for some time. As with RE elements, thorium improves casting properties and alloys based on the Mg-Th system are also weldable.

Ternary Mg-Th-Zr alloys, e.g. HK31A (Mg-3Th-0.7Zr), are commercially available and, in the as-cast condition, have microstructures similar to the Mg-RE-Zr alloys. The alloys are normally given a T6 ageing treatment and investigations of precipitation also suggest that similarities exist between the ageing processes of the two types of alloys. Although GP zones are likely to occur at low temperatures, these have not been detected and the precipitation sequence has been described as:

$$SSSS \rightarrow \beta'' \rightarrow \beta' \rightarrow \beta(Mg_{23}Th_6)$$

The phase β'' again has an ordered DO_{19} structure and occurs as thin discs which are coherent with the $\{10\overline{1}0\}$ and probably $\{11\overline{2}0\}$ planes of the matrix. There is dispute as to whether the formula of this compound is $MgTh_3$ or Mg_3Th, although the latter would provide the low energy

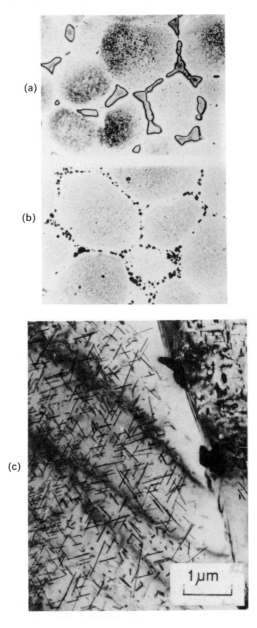

Fig. 5.11 Effect of hydriding on the microstructure of the alloy ZE63: a, as-cast microstructure; b, alloy heat treated in hydrogen atmosphere at 480° C and given T6 temper (courtesy Magnesium Elektron Ltd). ×300; c, thin-foil electron micrograph showing massive RE hydrides in the grain boundary and needles within the grains which are probably ZrH_2 (courtesy K. J. Gradwell). ×13000

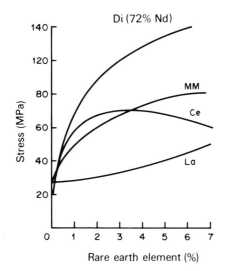

Fig. 5.12 Effect of various rare earth metals on the stress for 0.5% extension in 100 h at 205°C. Fully heat-treated Mg-RE alloys (from Leontis, T. E., *TAIMME*, **85**, 968, 1949)

Table 5.4 Tensile properties of cast test bars of the alloys ZK61 and ZE62 after solution treatment in an atmosphere of (a) SO_2 and (b) wet H_2 (from Fisher, P. A. *et al.*, *Foundry*, **95**(8), 68, 1967)

	Heat treatment: Solution treatment 24 h 500°C; Aged 64 h 125°C	Tensile properties		
Alloy		0.2% proof stress (MPa)	Tensile strength (MPa)	Elongation (% in 5 cm)
ZK61 (Mg-6Zn-0.7Zr)	SO_2	186	271	3.5
	H_2	183	252	2.5
ZE62 (ZK61 +2%RE)	SO_2	94	173	3.8
	H_2	181	306	12.0

interfaces with the matrix proposed in Fig. 5.2. This phase may transform directly to the equilibrium β-phase although two semi-coherent polymorphs β'_1 and β'_2 have been detected which form on dislocation lines in cold-worked alloys. The presence of zirconium indirectly favours the formation of one or both of these phases as they form on dislocations generated around zirconium-containing compounds. All these phases, as well as the equilibrium precipitate, appear to be resistant to coarsening at temperatures up to 350°C.

Again, in parallel with alloys based on the Mg-RE system, thorium-containing alloys have been developed to which zinc has been added, e.g. HZ32A (Mg-3Th-2.2Zn-0.7Zr) and ZH62A (Mg-5.7Zn-1.8Th-0.7Zr). The presence of zinc further increases creep strength (Fig. 5.13) and a Th:Zn ratio of 1.4:1 appears to be optimal in this regard. This ratio corresponds to a microstructure in which an acicular phase predominates in the grain boundaries and the good creep properties of the alloy HZ32A are attributed, at least in part, to the presence of this phase. Alloy ZH62A is noted for its relatively high strength at room temperature. The possible influence of zinc on precipitation in Mg-Th alloys is unknown.

5.5.4 Alloys based on the magnesium-silver system

The importance of this class of alloys stems from the discovery by Payne and Bailey that the relatively low tensile properties of age-hardened Mg-RE-Zr alloys could be much increased by the addition of silver. Room temperature tensile properties were obtained which were similar to those of the high-strength Mg-Zn-Zr alloys, such as ZK51, with the experimental

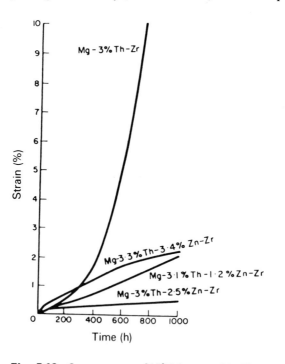

Fig. 5.13 Creep curves at 315°C for some Mg-Th-Zr and Mg-Th-Zn-Zr sand cast alloys. Note the beneficial effect of zinc on long term creep resistance and that there is evidently an optimum zinc content (from Ball, C. J., *TAIMME*, **197**, 924, 1953)

alloys having superior casting or welding characteristics. Substitution of normal cerium-rich misch metal with didymium misch metal (average composition 80 % Nd, 16 % Pr, 2 % Gd, 2 % others) gave a further increase in strength which was attributed to the presence of neodymium. Several commercial compositions were subsequently developed having tensile properties which exceeded those of any other magnesium alloys at temperatures up to 250° C, and which were comparable with the high-strength aluminium casting alloys.

The most widely used alloy has been QE22 (Mg-2.5Ag-2RE (Nd)-0.7Zr) for which the optimal heat treatment is: solution treatment for 4–8 h at 525° C, cold-water quench, age 8–16 h at 200° C. If these alloys contain less than 2 % silver the precipitation process appears to be similar to that occurring in Mg-RE alloys and involves the formation of Mg-Nd precipitates. However, for higher silver contents, this element apparently modifies precipitation and increases the volume fraction of particles that are formed (Fig. 5.14). Two independent precipitation processes have been reported, both of which lead ultimately to the formation of an equilibrium phase of probable composition $Mg_{12}Nd_2Ag$ (Table 5.3). The presence of a precipitate with the DO_{19} structure has not been confirmed although the phase designated γ has characteristics which suggest that it may be such a phase. Maximum age hardening and creep resistance are associated with the presence of γ- and β-precipitates in the microstructure.

Alloy QE22 has been used for a number of aerospace applications, e.g. aircraft landing wheels, gearbox housings and helicopter rotor fittings. Its

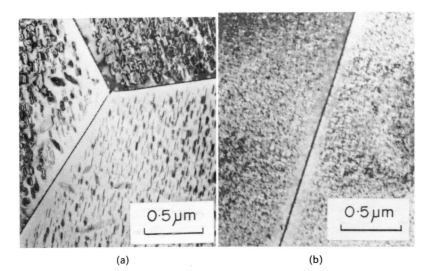

(a)   (b)

Fig 5.14   Thin-foil electron micrographs showing the effect of silver on precipitate size in Mg-RE(Nd)-Zr alloys aged to peak hardness at 200° C: a, EK21 (Mg-2.5RE(Nd)-0.7Zr); b, QE22 (Mg-2.5Ag-2RE(Nd)-0.7Zr) (courtesy K. J. Gradwell). ×32 000

superior tensile properties over most magnesium alloys are maintained to 250° C (Fig. 5.7) although it is only considered to be resistant to creep at temperatures up to 200° C (Figs. 5.7 and 5.8). The alloys are relatively expensive and attempts have been made to replace at least some of the silver with copper. This work has met with some success despite the relatively low solubility of copper in magnesium (maximum of 0.55 % at the eutectic temperature) although no practical alloys have so far been produced.

Developments in the aerospace industries provide a continuing requirement for higher operating temperatures, particularly in gas turbine engines, and recent work has been directed at extending the elevated temperature stability of the alloy QE22. The effects of adding thorium were investigated (see Section 5.5.3) and a new alloy QH21 has been developed in which part of the RE(Nd) addition is replaced by this element. The nominal composition of QH21 is Mg-2.5Ag-1Th-1RE(Nd)-0.7Zr and the stress-rupture characteristics of the two alloys at 250° C are shown in Figs. 5.15 and 5.16. Although thorium has the disadvantage of reducing the solid solubility of neodymium, the precipitate which forms on ageing is refined in size and has a slower rate of coarsening.

Because relatively little is known about some areas of the alloying behaviour of magnesium, it is to be expected that more new compositions will be developed having improved properties. For example, recent work by Borradaile has suggested that certain alloys containing yttrium may

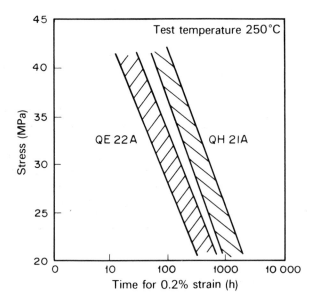

Fig. 5.15 Stress-rupture characteristics of the silver-containing alloys QH21 and QE22A at 250° C (courtesy Magnesium Elektron Ltd)

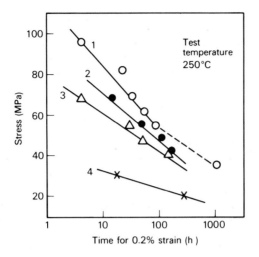

Fig. 5.16 Effect of stress on time to produce 0.2% strain at 250°C for the alloys QE22, QH21 and two experimental compositions containing yttrium. 1 = Mg-2.5Ag-2RE(Nd)-0.6Zr-4Y, 2 = Mg-2.5Ag-2RE(Nd)-0.6Zr-2Y, 3 = QH21 (Mg-2.5Ag-1RE(Nd)-1Th-0.6Zr), 4 = QE22 (Mg-2.5Ag-2RE(Nd)-0.6Zr) (from Borradaile, J. B., PhD Thesis, University of Liverpool, 1978)

have creep properties better than either of the alloys QE22 and QH21. Yttrium has a maximum solid solubility in magnesium of 12.5% which is significantly greater than that of thorium or the rare earth elements. Two experimental alloys have been studied, based on Mg-2.5Ag-2RE(Nd)-0.6Zr, with additions of 2% and 4% yttrium respectively. Room temperature tensile properties are comparable with those of alloys QE22 and QH21, but curves showing the effect of stress on time to produce 0.2% strain at 250°C (Fig. 5.16) have indicated that the yttrium-containing alloys show superior creep resistance. This effect has been attributed primarily to solid solution strengthening in the alloy having 2% yttrium, and to modifications to the dispersion of precipitates in the alloy containing 4% yttrium. However, from the practical viewpoint, it should be noted that yttrium is both costly and difficult to add to molten magnesium.

5.6 Wrought magnesium alloys

5.6.1 Introduction

The hexagonal crystal structure of magnesium places limitations on the amount of deformation that can be tolerated, particularly at low temperatures. At room temperature, deformation occurs mainly by slip on the basal planes in the close packed $<11\bar{2}0>$ directions and by twinning on the pyramidal $\{10\bar{1}2\}$ planes (Fig. 5.1). With stresses parallel to the basal planes, twinning of this type is only possible in compression whereas, with stresses perpendicular to the basal planes, it is only possible in tension. Above about 250°C, additional pyramidal $\{10\bar{1}1\}$ slip planes become operative so that deformation becomes much easier and twinning is less important. Production of wrought magnesium alloy products is, therefore, normally carried out by hot-working.

Wrought materials are produced mainly by extrusion, rolling and press forging at temperatures in the range 300 to 500°C and some general remarks can be made concerning the relative properties in various directions in the final products:
(i) Since the elastic modulus does not show much variation in different directions of the hexagonal magnesium crystal, preferred orientation has relatively little effect upon the modulus of wrought products.
(ii) Extrusion at relatively low temperatures tends to orient the basal planes and also the $<10\bar{1}0>$ directions approximately parallel to the direction of extrusion. Rolling tends to orient the basal planes parallel to the surface of sheet with the $<10\bar{1}0>$ directions in the rolling direction.
(iii) Because twinning readily occurs when compressive stresses are parallel to the basal plane, wrought magnesium alloys tend to show lower values of longitudinal proof stress in compression than in tension. The ratio may lie between 0.5 and 0.7 and, since the design of lightweight structures involves buckling properties which, in turn, are strongly dependent on compressive strength, the ratio is an important characteristic of wrought magnesium alloys. The value varies with different alloys and is increased by promoting fine grain size because the contribution of grain boundaries to overall strength becomes proportionally greater.
(iv) Strengthening of wrought products by cold-reeling in which alternate tension and compression occurs can cause extensive twinning through compression, with a marked reduction in tensile properties.

As with cast alloys, the wrought alloys may be divided into two groups according to whether or not they contain zirconium. However, it is proposed here to consider the alloys with regard to the form of the wrought product. Compositions and mechanical properties are summarized in Table 5.2. Discussion of the ageing behaviour is only included where the compositions differ significantly from the cast alloys.

5.6.2 Sheet and plate alloys

The early sheet alloys were AZ31 (Mg-3A1-1Zn-0.3Mn), which is still the most widely used magnesium alloy for applications at room or slightly elevated temperatures, and the now little used alloy M1A (Mg-1.5Mn). AZ31 is strengthened by strain hardening and is weldable, although weldments should be stress-relieved to minimize susceptibility to stress-corrosion cracking. Higher room temperature properties can be obtained with the British alloy ZK31 (Mg-3Zn-0.7Zr) but weldability is limited. Two lower strength alloys ZM21 (Mg-2Zn-1Mn) and ZE10 (Mg-1.2Zn-0.2RE) are weldable and do not require stress-relieving. ZE10 has the highest toughness of any magnesium sheet alloy.

The Mg-Li system has attracted attention as a basis for very lightweight sheet and plate. Lithium with a relative density of 0.53 is the lightest of all metals and the Mg-Li phase diagram (Fig. 5.17) shows this element to have extensive solid solubility in magnesium. Moreover, only about 11% lithium is needed to form a new β-phase, which has a body-centred cubic (bcc) structure, thereby offering the prospect of extensive cold-formability. Finally, the slope of the $\alpha + \beta/\beta$ phase boundary suggested that selected compositions may show age hardening. Early work on binary alloys revealed that traces of sodium caused grain boundary embrittlement but this problem was overcome with the availability of high-purity lithium. A second difficulty was that the binary alloys became unstable and over-aged at slightly elevated temperatures (50–70° C) resulting in excessive creep under relatively low loads. Greater stability has since been achieved by adding other elements and one composition LA141 (Mg-14Li-1Al), which is weldable, has been used for armour plate and for aerospace components. It has a relative density of only 1.35 and a specific modulus or stiffness (E/d) which is second only to beryllium (Chapter 1). Elevated temperature stability is further increased by the addition of 0.5% silicon.

Thorium-containing wrought alloys have also been developed for elevated temperature applications. The first of these was HK31 (Mg-3Th-0.6Zr) which requires the full T6 heat treatment to develop maximum creep strength. In a later alloy, HM21 (Mg-2Th-0.6Mn), cold-working prior to ageing (T8 temper) increases strength at temperatures up to 350° C. The

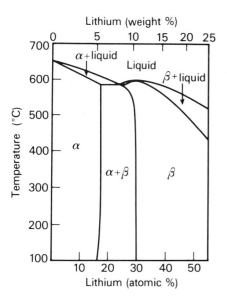

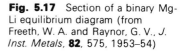
Fig. 5.17 Section of a binary Mg-Li equilibrium diagram (from Freeth, W. A. and Raynor, G. V., *J. Inst. Metals*, **82**, 575, 1953–54)

British alloy ZTY (Mg-0.75Th-0.5Zn-0.6Zr) has creep properties comparable with HK31 and HM21, but has the advantage that no heat treatment is needed. All three alloys are weldable.

Only limited cold-forming can be carried out with sheet alloys and typical minimum bend radii vary from 5 to $10T$, where $T =$ thickness of sheet, for annealed material, and from 10 to $20T$ for the hard-rolled condition. Thus, for even simple operations, hot-forming within the temperature range 230 to 350° C is preferred. Under these conditions sheet can be formed by pressing, deep drawing, spinning and other methods using relatively low-powered machinery.

5.6.3 Extrusion alloys

A wide range of Mg-Al-Zn extrusion alloys are used with aluminium contents between 1 and 8%, the strongest alloy AZ81 (Mg-8Al-1Zn-0.7Mn) showing some response to age hardening if heat treated after fabrication. AZ61 is widely used in Britain as a general purpose light alloy.

One composition was specifically developed as a canning material for use in the British gas cooled, Magnox nuclear reactors. This has the composition Mg-0.8Al-0.005Be and the fuel element cans are impact extruded with integral cooling fins, as shown in Fig. 5.18, or machined from a finned extrusion. The spiral shape is obtained by hot-twisting after extrusion. The selection of a magnesium alloy was made because the metal has a relatively low capture cross-section for thermal neutrons (0.059 barns), is resistant to creep and corrosion by the carbon dioxide coolant at the operating temperatures (180 to 420° C) and, contrary to aluminium, does not react with the uranium fuel. The addition of aluminium provides some solid solution strengthening whereas the trace amount of beryllium improves oxidation resistance.

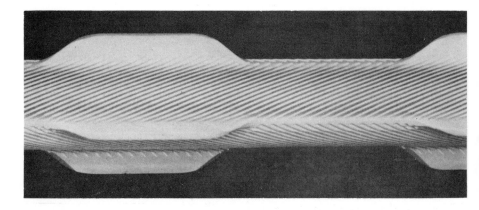

Fig. 5.18   Part of extruded magnesium alloy fuel can from a British Magnox nuclear reactor

The alloy ZK61 (Mg-6Zn-0.7Zr), which is normally aged after extrusion, develops the highest room temperature yield strength of the more commonly used wrought magnesium alloys. It also offers the advantage that tensile and compressive yield strengths are closely matched. The lower zinc alloys ZK21 and ZM21 (Mg-2Zn-1Mn) are widely used where higher extrusion rates are desired. The highest strength recorded for a wrought magnesium alloy was for the composition Mg-6Zn-1.2Mn (ZM61) in the form of heat treated extruded bar. The alloy shows a high response to age hardening, and solution treatment at 420° C, water quenching and duplex ageing below and above the GP zones solvus (24 h at 90° C and 16 h at 180° C) results in the following tensile properties: 0.1 % proof stress 340 MPa; tensile strength 385 MPa; with an elongation of 8 %. On a strength/weight basis, this alloy has properties comparable with some of the strongest wrought aluminium alloys.

For elevated temperatures, the thorium-containing alloys HK31 and HM31 (Mg-3Th-1Mn) are used for components which are creep-resistant to 350° C, with the latter alloy being capable of withstanding short-term exposures at temperatures up to 425° C.

5.6.4 Forging alloys

Forgings represent a relatively small part of wrought magnesium products and are generally used where a lightweight part of intricate shape is required having strengths higher than can be achieved with castings. Thus forgings tend to be made from the higher strength alloys AZ80 or ZK60 for use at room temperature, and HK31 or HM21 for elevated temperature applications. Press forging is more common than hammer forging and it is common practice to pre-extrude the forging blanks to refine the microstructure.

5.7 Electrochemical aspects

5.7.1 Corrosion and protection

Magnesium has a normal electrode potential at 25° C of − 2.30V, with respect to the hydrogen electrode potential taken as zero, which places it high in the electrochemical series. However, its solution potential is lower, e.g. − 1.7 V in dilute chloride solution with respect to a normal calomel electrode, due to polarization of the surface with a film of Mg (OH)$_2$. The oxide film on magnesium offers considerable surface protection in rural and most industrial environments and the corrosion rate of magnesium lies between aluminium and mild steel (Table 5.5). Tarnishing occurs readily and some general surface roughening may take place after long periods but, unlike some aluminium alloys, magnesium and its alloys are virtually immune from intercrystalline attack.

The early magnesium alloys suffered rapid attack in moist conditions, due mainly to the presence of impurities, notably iron and nickel. Iron has a

Table 5.5   Results of $2\frac{1}{2}$-year exposure tests (from *Metals Handbook*, Volume 1, 8th edition, American Society of Metals, Cleveland 1961)

Material	Corrosion rate (mm yr^{-1})	Tensile strength after $2\frac{1}{2}$ years (% loss)
Marine atmosphere		
Aluminium alloy 2024	0.002	2.5
Magnesium alloy AZ31	0.028	7.4
Mild steel	0.236	75.4
Industrial atmosphere		
Aluminium alloy 2024	0.003	1.5
Magnesium alloy AZ31	0.044	11.2
Mild steel	0.040	11.9
Rural atmosphere		
Aluminium alloy 2024	0.000	0.4
Magnesium alloy AZ31	0.021	5.9
Mild steel	0.024	7.5

particularly deleterious effect on corrosion if the level exceeds 0.017% in certain alloys. Even less iron can be tolerated in Mg-Al or Mg-Al-Zn alloys. Nickel is more harmful than iron and these elements, or the compounds they form, act as minute cathodes in the presence of a corroding medium creating micro-cells with the relatively anodic magnesium matrix. It is now possible to have a closer control on impurity limits during magnesium production and small additions of manganese also minimize the adverse effects of these impurities in zirconium-free alloys. Zirconium has a similar effect on those alloys to which it is added. Special care must be exercised to ensure that iron scale does not adhere to the surface of wrought products during fabrication. Similarly, emery grit or iron particles used to shot blast foundry sands from castings can adhere and act as cathodes.

Magnesium is readily attacked by all mineral acids except chromic and hydrofluoric acids, the latter actually producing a protective film of MgF_2 which prevents attack by most other acids. In contrast, magnesium is very resistant to corrosion by alkalis if the pH exceeds 10.5 which corresponds to that of a saturated $Mg(OH)_2$ solution. Chloride ions promote rapid attack of magnesium in aqueous solutions as do sulphate and nitrate ions, whereas soluble fluorides are chemically inert. With organic solutions, methyl alcohol and glycol attack magnesium whereas ethyl alcohol, methylated spirits, oils and degreasing agents are inert.

The corrosion behaviour of alloys varies with composition. Where alloying elements form grain boundary phases, as is generally the case in casting alloys, corrosion rates are likely to be greater than those occurring with pure magnesium.

Some magnesium alloys may be susceptible to stress-corrosion cracking (SCC) if subjected to tensile stress and exposed to distilled water, dilute chlorides and some other solutions. Cracking is primarily transgranular and appears to show comparatively little relationship to microstructural features such as slip or twinning planes. Mg-Al alloys have the greatest susceptibility, whereas in the zirconium-containing alloys SCC only occurs at stresses approaching the yield stress of the alloy concerned and is not regarded as a serious problem. Wrought products are more likely to undergo SCC than castings and it is desirable to stress-relieve if a component is to be exposed to potential corrodents.

It is common practice to protect the surface of magnesium and its alloys and such protection is essential where contact with other metals may lead to galvanic corrosion. The methods available for magnesium are:

(i) Fluoride anodizing – this involves alternating current anodizing at up to 120 V in a bath of 25 % ammonium bifluoride which removes surface impurities and produces a thin, pearly white film of MgF_2. This film is normally stripped in boiling chromic acid before further treatment as it gives poor adhesion to organic treatments.

(ii) Chemical treatments involving pickling and conversion of the oxide coating – components are dipped in chromate solutions which clean and passivate the surface to some extent through formation of a film of $Mg(OH)_2$ and a chromium compound. Such films have only slight protective value, but form a good base for subsequent organic coatings.

(iii) Electrolytic anodizing, including proprietary treatments that deposit a hard ceramic-like coating which offers some abrasion resistance in addition to corrosion protection, e.g. Dow 17, HEA, and MGZ treatments – such films are very porous and provide little protection in the unsealed state but they may be sealed by immersion in a solution of hot dilute sodium dichromate and ammonium bifluoride, followed by draining and drying. A better method is to impregnate with a high temperature curing epoxy resin (see (iv)). Resin-sealed anodic films offer very high resistance to both corrosion and abrasion, and, in some instances, can even be honed to provide a bearing surface.

(iv) Sealing with epoxy resins – in this case, the component is heated to 200 to 220° C to remove moisture, cooled to approximately 60° C, and dipped in the resin solution. After removal from this solution, draining and air-drying to evaporate solvents, the component is baked at 200 to 220° C to polymerize the resin. Heat treatment may be repeated once or twice to build up the desired coating thickness which is commonly 0.0025 mm.

(v) Standard paint finishes – the surface of the component should be prepared as in (i) to (iv), after which it is preferable to apply a chromate-inhibited primer followed by good quality top coat.

(vi) Vitreous enamelling – such treatments may be applied to alloys which do not possess too low a solidus temperature. Surface preparation involves dipping in a chromate solution before applying the frit.

(vii) Electroplating – several stages of surface cleaning and the application of pre-treatments, such as a zinc conversion coating, are required before depositing chromium, nickel or some other metal.

Magnesium alloy components for aerospace applications require maximum protection; schemes involving chemical cleaning by fluoride anodizing, pre-treatment by chromating or anodizing, sealing with epoxy

resin, followed by chromate primer and top coat are sometimes
mandatory.

5.7.2 Cathodic protection

Due to its very active electrode potential, magnesium and its alloys can be
used to protect many other structural materials from corrosion when
connected to them in a closed electrical circuit. Magnesium acts as an
anode and is consumed sacrificially, thereby offering protection to metals
such as steel. Magnesium metal and, more commonly, the alloys AZ63 and
M1A (Mg-1.5Mn) which offer higher relative voltages are used for this
purpose. Examples of areas where cathodic protection is used are ships'
hulls, pipelines and steel piles. It should be noted, however, that magnesium
and its alloys are not used to protect oil rigs because of the potential
incendive sparking risk.

5.8 Fabrication of components

5.8.1 Machining

Magnesium and its alloys are the most machinable of all structural
materials. This applies with respect to depth of cut, speed of machining,
tool wear and relative amounts of power required for the equipment being
used (Table 5.6). Magnesium is normally machined dry but, where very
high cutting speeds are involved and there is a possibility of igniting fine
turnings, it may be necessary to employ a coolant. For this purpose,
mineral oils must be used because water-based coolants may react
chemically with the swarf. Good tool life is experienced providing cutting
edges are kept sharp and generous rake clearance angles (usually $7°$
minimum) are used. Sharp tools also reduce the possibility of fires due to
frictional heat.

Magnesium alloys can be chemically machined or milled by pickling in
$5\% H_2SO_4$ or in dilute solutions of HNO_3 or HCl. Some alloys also lend
themselves to contour etching and AZ31 is widely used for the production
of printing plates.

Table 5.6 Comparative machinability of metals (from *Machining*, Magnesium
Elektron Limited Handbook)

Metal	Relative power required •	Rough turning speeds (ms^{-1})	Drilling speeds (5–10 mm drill) (ms^{-1})
Magnesium	1	up to 20	2.5–8.5
Aluminium	1.8	1.25–12.5	1–6.5
Cast iron	3.5	0.5–1.5	0.2–0.65
Mild steel	6.3	0.65–3.3	0.25–0.5
Stainless steel	10.0	0.3–1.5	0.1–0.35

• 1 = lowest

5.8.2 Joining

Early magnesium alloys were gas welded with an oxyacetylene torch and required careful fluxing to minimize oxidation. Apart from the normal difficulties associated with such a process, extensive corrosion of welds was common when the flux was incompletely removed by the cleaning methods applied. Now virtually all magnesium welding is done with the inert gas shielded tungsten arc (TIG) or consumable electrode (MIG) processes. Spot welding and other forms of resistance welding can be carried out although they are comparatively little used.

Both cast and wrought products can be welded and the weldability of different alloys has already been discussed and compared in Tables 5.1 and 5.2. In general, filler rods of the same composition as the parent alloy are desirable although the use of a more highly alloyed rod with lower melting point and wider freezing range is sometimes beneficial to minimize cracking. Castings are often preheated to 250 to 300°C to reduce weld cracking during solidification, and stress-relieving may be desirable after welding is completed.

The design of mechanical joints in magnesium and its alloys is qualified by the vital consideration of galvanic corrosion. This problem precludes direct contact with most other metals and special coatings or insulating materials must be used as separating media. Care must also be taken in the design of joints to avoid crevices, grooves and such like, where water and other corrosive materials can collect.

Further reading

Emley, E. F., *Principles of Magnesium Technology*, Pergamon, London, 1966

Roberts, C. S., *Magnesium and its Alloys*, Wiley, New York, 1960

Raynor, G. V., *The Physical Metallurgy of Magnesium and its Alloys*, Pergamon, London, 1959

Hallowell, J. B. and Ogden, H. R., An introduction to magnesium alloys, *Defence Metals Information Center DMIC Report 206*, Battelle Memorial Institute, Columbus, 1964

Raynor, G. V., Constitution of ternary and some more complex alloys of magnesium, *Int. Met. Rev.*, **22**, 65, 1977

Stratford, D. J. and Beckley, L., Precipitation processes in Mg-Th, Mg-Th-Mn, Mg-Mn and Mg-Zr alloys, *Met. Sci. J.*, **6**, 83, 1972

Fisher, P. A. *et al.*, New high strength magnesium casting alloys for aerospace applications, *Foundry*, **95**, No. 8, 68, 1967

Fletcher, S. J., Magnesium and its applications, *Eng. Mater. and Design*, **17**, 13, 1973

6
Titanium alloys

6.1 Introduction

Stimulus for the development of titanium alloys during the past 30 years came initially from the aerospace industries when there was a critical need for new materials with higher strength:weight ratios at elevated temperatures. As mentioned in Chapter 1, the high melting point of titanium (1678°C) was taken as a strong indication that the alloys would show good creep strengths over a wide temperature range. Although subsequent investigations revealed that this temperature range was narrower than expected, titanium alloys now occupy a critical position in the materials inventory of the aerospace industries (see Fig. 1.3) and 80–90 % of titanium is used in this way. More recently the importance of these alloys as corrosion resistant materials has been appreciated by the chemical industry as well as by the medical profession which uses titanium alloy prostheses for implanting in the human body. It is proposed to consider the alloys with respect to these applications and to concentrate on wrought products as titanium alloy castings amount to less than 1 % of titanium metal.

Titanium has a number of features that distinguish it from the other light metals and which make its physical metallurgy both complex and interesting:

(i) At 882°C, titanium undergoes an allotropic transformation from a low temperature, hexagonal close-packed structure (α) to a body-centred cubic (β) phase that remains stable up to the melting point. This transformation offers the prospect of having alloys with α, β or mixed α/β microstructures and, by analogy with steels, the possibility of using heat treatment to extend further the range of phases that may be formed.

(ii) Titanium is a transition metal with an incomplete d shell in its electronic structure which enables it to form solid solutions with most substitutional elements having a size factor within $\pm 20\%$.

(iii) Titanium and its alloys react with several interstitial elements including the gases oxygen, nitrogen and hydrogen, and such reactions may occur at temperatures well below the respective melting points.

(iv) In its reactions with other elements, titanium may form solid solutions and compounds with metallic, covalent or ionic bonding.

Alloying of titanium is dominated by the ability of elements to stabilize either of the α- or the β-phases. This behaviour, in turn, is related to the

162

number of bonding electrons, i.e. the group number, of the element concerned: alloying elements with electron/atom ratios of less than 4 stabilize the α-phase, elements with a ratio of 4 are neutral, and elements with ratios greater than 4 are β-stabilizing.

6.1.1 Classification of titanium alloys

Titanium alloy phase diagrams are often complex and many are unavailable. However, the titanium-rich sections of pseudo-binary systems enables them to be classified into three simple types, as shown in Fig. 6.1. Elements that dissolve preferentially in the α-phase expand this field thereby raising the α/β transus (Fig. 6.1a) and, of the comparatively few elements that behave in this way, aluminium and oxygen are the most important. Zirconium, tin and silicon are regarded as neutral in their effect on either phase. Elements which depress the α/β transus and stabilize the β-phase may be classified in two groups: those which form binary systems of the β-isomorphous type (Fig. 6.1b) and those which favour formation of a β-eutectoid (Fig. 6.1c). It should be noted, however, that the eutectoid reactions in a number of alloys are very sluggish so that, in practice, the alloys tend to behave as if this reaction did not occur. Examples are the binary systems Ti-Fe and Ti-Mn and these alloys behave as if they conformed to the β-isomorphous phase diagram, hence the arrows shown in Fig. 6.1c.

The main elements that promote the three types of binary phase diagrams are also given in Fig. 6.1. It will be noted that the interstitial elements also exert stabilizing effects: oxygen, nitrogen and carbon favouring the α-phase and hydrogen promoting the β-phase. Of the

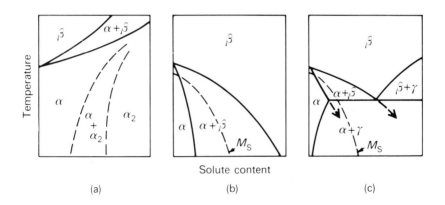

Fig. 6.1 Basic types of phase diagrams for titanium alloys. The dotted phase boundaries in (a) refer specifically to the Ti-Al system. The dotted lines in (b) and (c) show the martensite start (M_s) temperatures. Alloying elements favouring the different types of phase diagrams are: a, Al, O, N, C, Ga; b, Mo, W, V, Ta; c, Cu, Mn, Cr, Fe, Ni, Co, H

substitutional elements that stabilize the β-phase, molybdenum and tungsten have the greatest effects although the latter element is little used because of its high density and problems of segregation during alloy preparation. Vanadium is another common β-stabilizer although it is less effective than molybdenum in the higher temperature ranges.

It is customary to classify titanium alloys into three main groups designated α, α + β and β which will each be considered in Sections 6.2 to 6.4. The compositions and a selection of properties of representative commercial alloys in each group are listed in Table 6.1. In addition, the creep characteristics of a number of these alloys are shown in Fig. 6.2 because this property has dominated much alloy development. Both Table 6.1 and Fig. 6.2 should be consulted in conjunction with the foregoing discussions in which special consideration is given to commercial compositions and the roles of particular alloying elements.

6.1.2 Basic principles of heat treatment

Although the first alloys that will be discussed (α-alloys) show little response to heat treatment, it is desirable to examine the general principles that are involved even though they relate mainly to the α/β group. This is possible by considering the effects of alloy content on the β to α transformation in a typical binary β-isomorphous system, as shown in Fig. 6.3. Also included is

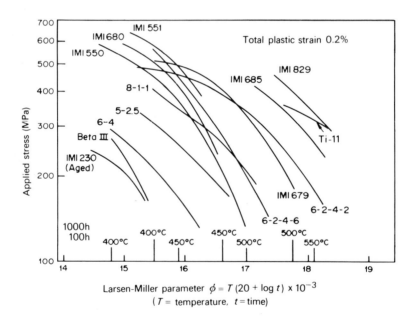

Fig. 6.2 Creep curves of some commercial titanium alloys (courtesy IMI Titanium)

Table 6.1 Compositions, relative densities and typical room temperature tensile properties of selected wrought titanium alloys

Common designations	Al	Sn	Zr	Mo	V	Si	Other	Relative density	Condition	0.2% proof stress (MPa)	Tensile strength (MPa)	Elongation (%)
α-alloys												
CP Ti 99.5% IMI 115, Ti-35A							0	4.51	Annealed 675°C	170	240	25
CP Ti 99.0% IMI 155, Ti-75A							0	4.51	Annealed 675°C	480	550	15
IMI 260							0.2 Pd	4.51	Annealed 675°C	315	425	25
IMI 317	5	2.5						4.46	Annealed 900°C	800	860	15
IMI 230							2.5 Cu	4.56	ST (α)●, duplex aged 400 and 475°C	630	790	24
Near-α alloys												
8-1-1	8			1	1			4.37	Annealed ▲780°C	950	990	15
IMI 679	2.25	11	5	1		0.25		4.82	ST (α+β) aged ■500°C	990	1100	15
IMI 685	6		5	0.5		0.25		4.49	ST (β) aged 550°C	900	1020	12
6-2-4-2	6	2	4	2		0.1		4.54	ST (α+β) annealed 590°C	960	1030	15
Ti-11	6	2	1.5	1		0.1	0.35 Bi	4.45	ST (β) aged 700°C	850	940	15
IMI 829	5.5	3.5	3	0.3		0.3	1 Nb	4.61	ST (β) aged 625°C	–	–	–
α/β alloys												
IMI 318, 6-4	6				4			4.46	Annealed 700°C	925	990	14
IMI 550	4	2		4		0.5		4.60	ST (α+β) aged 500°C	1100	1170	10
IMI 680	2.25	11		4		0.2		4.86	ST (α+β) aged 500°C	1000	1100	14
6-6-2	6	2		6	6		0.7 (Fe, Cu)	4.54	ST (α+β) aged 500°C	1190	1310	15
6-2-4-6	6	2	4	6				4.68	ST (α+β) aged 550°C	1170	1275	10
									ST (α+β) annealed 590°C	≈1170	1270	10
IMI 551	4	4	4	4		0.5		4.62	ST (α+β) aged 500°C	1200	1310	13
Ti-8 Mn							8 Mn	4.72	Annealed 700°C	860	945	15
β-alloys												
13-11-3	3			11.5	13		11 Cr	4.87	ST (β) aged 480°C	1200	1280	8
Beta III		4.5	6	11.5				5.07	ST (β) duplex aged 480 and 600°C	1315	1390	10
8-8-2-3	3	2	11	8	8		2 Fe	4.85	ST (β) aged 580°C	1240	1310	8
Transage 129	2	2	11	4	11			4.81	ST (β) aged 580°C	1280	1400	6
Beta C	3		4	4	8		6 Cr	4.82	ST (β) aged 540°C	1130	1225	10

● ST (α), ST (α+β), ST (β) correspond to solution treatment in the α, α+β, and β-phase fields respectively

▲ ■ Annealing treatments normally involve shorter times than ageing treatments

a schematic diagram which depicts trends in tensile strength with respect to alloy content resulting from different heat treatment procedures.

It will be seen that the strength of annealed alloys increases gradually and linearly as alloy content, or percentage of β-phase, increases. It should be noted that the β-phase in these alloys does not transform during cooling to room temperature. However, for alloys quenched from the β-phase field, a more complex relationship exists between strength and composition which is dependent upon the transformation of β to the martensitic form of the α-phase, designated α'. For low concentrations of solute, some strengthening occurs as a result of this transformation but the effect is much less than that traditionally found for martensitic reactions in ferrous materials. Moreover little change occurs when martensitic α' is tempered or aged. The maximum strength obtainable from this β to α' transformation occurs at a

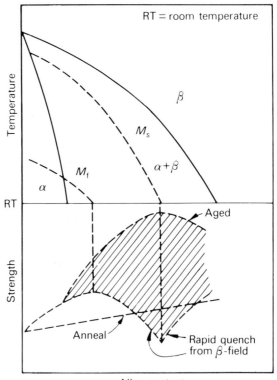

Fig. 6.3 Schematic diagram for the heat treatment of β-isomorphous titanium alloys (from Morton, P. H., *Rosenhain Centenary Conference on the Contribution of Metallurgy to Engineering Practice*, The Royal Society, London, 1976)

composition for which the martensite finish (M_f) temperature corresponds to room temperature.

Increasing the solute content above this level results in a progressive increase in the amount of metastable β that is retained on quenching from the β or $\alpha + \beta$ phase fields, and there is a gradual decrease in strength of quenched alloys to a minimum value at a composition at which the martensite start (M_s) temperature occurs at room temperature, i.e. 100% metastable β. On the other hand, these compositions provide the maximum response to strengthening if the quenched alloys are aged to decompose the retained β (see shaded area in Fig. 6.3).

The formation and decomposition of martensitic α', and the ageing of alloys containing metastable β, both involve a number of complex reactions some of which have little significance in the heat treatment of commercial alloys. These reactions are discussed when considering appropriate groups of titanium alloys.

6.2 α-alloys

6.2.1 General

The main substitutional alloying elements which dissolve in the α-phase are the stabilizing elements aluminium and oxygen and the neutral elements tin and zirconium. All cause solid solution hardening and increase tensile strength by between 35 and 70 MPa for each one per cent of the added element. Oxygen and nitrogen, which are normally present as impurities, contribute to interstitial hardening and controlled amounts of oxygen are used to provide a specific range of strength levels in several grades of what is known as commercial-purity (CP) titanium. Small amounts of other elements can be present and α-alloys are often divided into three sub-groups depending upon whether they are entirely single phase α, contain up to 2% of β-stabilizing elements (near-α alloys) or respond to a conventional age-hardening reaction (Ti-Cu alloys with <2.5% Cu).

There is a practical limit to the amount of α-stabilizing elements that may be added to titanium since the alloys tend to embrittle because of an ordering reaction that occurs if the 'aluminium equivalent' exceeds about 9%. This quantity, or ordering parameter, may be calculated empirically from the composition by summing the weight percentage as follows:

$$Al + \frac{1}{3}Sn + \frac{1}{6}Zr + 10(O + C + 2N)$$

The ordering reaction has been widely studied, particularly in binary Ti-Al alloys, but it remains incompletely understood. In this system, what is known is that elevated temperature ageing of alloys with an aluminium content above 5 to 6% can lead to the formation of a finely dispersed, ordered phase (α_2) which is coherent with the lattice of the α-phase over a wide temperature range. The phase α_2 has the general formula Ti_3X and has the DO_{19} (hexagonal) crystal structure that was noted for precipitates formed in several magnesium alloys (Table 5.3). Continuing ageing gradually causes α_2 to coarsen. In other α-alloys such as Ti-Sn, and in more complex compositions, the misfit between α and α_2 is larger so that nucleation of α_2 becomes more difficult and it tends to form hetero-geneously. This non-uniform dispersion has a less deleterious effect on ductility. There is general agreement concerning the location of the $\alpha/(\alpha + \alpha_2)$ phase boundary up to 800°C, but uncertainty exists as to whether α_2 forms by peritectic reaction involving β and an intermediate compound TiAl, or by phase separation of α at higher temperatures (see Fig. 6.1a).

The poor ductility of α-alloys containing the α_2-phase has been a disappointment as the strengthening of the creep-resistant nickel alloys (superalloys) is based on microstructures containing a rather similar coherent phase $Ni_3(Ti, Al)$ or γ'. The only element reported to improve the ductility of α_2 in titanium alloys is gallium and, although some experimental gallium-containing alloys have been produced, the high cost of this element and problems with melting suggest that its commercial use is unlikely.

In the presence of hydrogen, titanium hydrides may form as long thin plates in pure and alloyed α-titanium. These hydrides may be important in fracture phenomena (see Section 6.7.2). They have the stochiometric composition TiH_2, although it has been reported that hydrogen:metal ratios may vary from 1.5 to 2. The basic structure is face-centred cubic but a tetragonal distortion develops at temperatures below 37°C. There is also evidence that a strain-induced hydride may form which has a body-centred cubic structure. A large volume expansion of as much as 18 % accompanies hydride formation and this may cause the generation of accommodation dislocations in the surrounding matrix.

6.2.2 Fully- α alloys

The only commonly used alloys in this group are the several grades of CP titanium, which are in effect Ti-O alloys, and the ternary composition Ti-5Al-2.5Sn. As the alloys are single phase, tensile strengths are relatively low although their high thermal stability leads to reasonable creep strengths in the upper temperature range (Fig. 6.2). They display good ductility down to very low temperatures and are readily weldable. As it is usually necessary to hot-work the alloys at temperatures below α/β transus in order to prevent excessive grain growth, formability is limited because of their hexagonal crystal structure and the fact that they exhibit a high rate of strain hardening. For this reason, Ti-5Al-2.5Sn has tended to be replaced by the age-hardenable Ti-Cu alloy which can be more easily fabricated after solution treatment but prior to ageing when it is relatively soft.

The mechanical properties of α-alloys are comparatively insensitive to microstructure although it is possible to obtain α in three different forms (Fig. 6.4):
(i) Equi-axed grains which are formed when the alloys are worked and annealed in the α-phase field (Fig. 6.4a). Grain sizes tend to be relatively

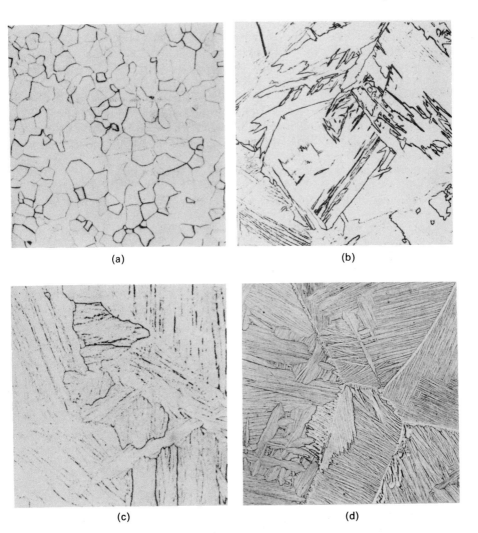

Fig. 6.4 Microstructure of CP titanium: a, annealed 1 h at 700° C showing equi-axed grains of α. ×100; b, quenched from β-phase field showing martensitic α'. ×150; c, air-cooled from the β-phase field showing Widmanstätten plates of α. ×100 (courtesy W. K. Boyd); d, near-α alloy IMI 685 air-cooled from the β-phase field showing a basket weave configuration of Widmanstätten plates of the α-phase delineated by small amounts of the β-phase (courtesy IMI Titanium). ×75

small because grain growth is inhibited due to the comparatively low temperatures that are involved and to the presence of impurities which pin grain boundaries. Yield strength at room temperature can be predicted from the Hall-Petch relationship, e.g. the equation for one grade of CP titanium (Ti-50A) is:

$$\sigma_{YS} = 231 + 10.54\,d^{-\frac{1}{2}}\ \text{(MPa)}$$

(ii) Quenching from the β-phase field produces the hexagonal martensitic phase α' in which the original β-grains remain clearly delineated. α' forms by a massive transformation, i.e. the martensite contains a high density of dislocations but few or no twins, and is composed of colonies of plates or laths separated by low angle boundaries. The transformation is characterised by a habit plane near $\{334\}_\beta$. There is negligible hardening associated with the production of α' martensite because the grain size is large and there is no supersaturation of the substitutional solute atoms. (iii) Slow cooling from β-phase field causes α to form as Widmanstätten plates (Fig. 6.4c). In high-purity alloys this structure is referred to as serrated α, whereas, if β-stabilizing elements or impurities such as hydrogen are present, the α-plates may be delineated producing a 'basket weave' effect (Fig. 6.4d).

The α-alloys that are cooled from the β-phase field exhibit lower values of tensile strength, room temperature fatigue strength and ductility, than those having an equi-axed grain structure. For low-cycle fatigue strength there is an empirical relation:

$$\text{fatigue strength at } 5 \times 10^4 \text{ cycles} \propto 0.1\,\% \text{ proof stress} \times \log R_A$$

where R_A = reduction in area in tensile test. This is a useful guide in rating alloys. On the other hand, cooling from the β-field leads to improved values of fracture toughness and higher creep resistance. These trends in mechanical properties which arise from the shape and size of the grains, and from the structure of the grain boundaries, are important as they are characteristic of many other titanium alloys.

CP titanium is the second most used titanium alloy which, for example, currently makes up some 20 % of all sales of titanium in the United States. It is available in most wrought forms and has extensive uses in both aerospace and general engineering, e.g. skin panels, fire walls and engine rings for aircraft, tubing for heat exchangers, valves and tanks for the chemical industry. In addition, a composition containing 0.2 % palladium (IMI 260) has been developed which has a particularly high resistance to corrosion.

Use of the alloy Ti-5Al-2.5Sn has declined in recent years as alloys with better forming properties and higher creep resistance have become available. One continuing application, however, has been cryogenic storage tanks for which the relatively high strength of titanium alloys at low temperatures is attractive. For this purpose a special grade which is low in interstitial elements (designated ELI) has been developed in the United States to increase the toughness of the alloy and it has been used to store liquid hydrogen ($-253°\text{C}$). Titanium alloy pressure vessels have become standard for fuel storage in a number of space vehicles as their specific strengths are approximately double those of aluminium alloys and stainless steels at such temperatures.

6.2.3 Near- α alloys

This class of forging alloys was developed to meet demands for higher operating temperatures in the compressor section of aircraft gas turbine engines as part of the continuing quest for improved performance and efficiency. They possess higher room temperature tensile strength than the fully-α alloys and show the greatest creep resistance of all titanium alloys at temperatures above approximately 400° C. Fig. 6.5 shows a forged compressor disc or wheel made from the near-α alloy IMI 685.

Bomberger has produced an empirical expression to denote those compositions giving maximum creep strength:

$$36 - 2.6(\%\ Al) - 1.1(\%\ Sn) - 0.7(\%\ Zr) - 27(\%\ Si) - 3(\%\ Mo\ \text{equivalent}) \leqslant 10$$

where $\%$ Mo equivalent $= \%Mo + 0.5(\%\ Nb) + 0.2(\%\ Ta) + 0.75(\%\ V) + 0.5(\%\ W) \leqslant 1.5$. In practice, the near-α alloys contain up to 2% β-stabilizing elements which both introduce small amounts of β-phase into the microstructure and improve forgeability. However, these additions are normally too small to provide significant strengthening through the decomposition of retained β (Fig. 6.3) and the improvement in mechanical properties arises mainly from the formation of martensitic α′ and from the manipulation of α/α′ microstructures.

Most near-α alloys are forged and heat treated in the α + β phase field so that primary α-grains are always present in the microstructure. More recently, improved creep performance has been achieved in special compositions by carrying out these operations at higher temperatures which places the alloys in the β-phase field and results in a change in the microstructure. It is now appropriate to consider the metallurgical synthesis of alloy compositions in each of these two categories.

Fig. 6.5 Forged compressor disc or wheel made from the near-α alloy IMI 685 (courtesy Rolls Royce Ltd)

Alloys heat treated in $\alpha + \beta$ phase field   One of the first alloys specifically designed to meet creep requirements was the composition Ti-11Sn-2.25Al-5Zr-1Mo-0.2Si (IMI 679). Development occurred in three well-defined stages. The first was to determine the maximum amounts of α-stabilizing elements that could be added without a severe loss of ductility and it was recognized that, although tin caused less solid solution hardening than aluminium at room temperature, it became a more effective strengthener as the temperature was raised. Tin also had the advantage that higher amounts could be tolerated without causing formation of the embrittling α_2-phase, although it was appreciated that such large amounts would prevent the alloy from being welded. In actual practice, the total content of tin and aluminium was limited by a tendency for the alloys to become susceptible to hydrogen embrittlement and a ternary composition Ti-11Sn-2.25Al was selected. Zirconium was added to provide further solid solution strengthening of the α-phase.

The second stage of development was to introduce a β-stabilizing element that would both promote some response to heat treatment and render the alloy more forgeable without adversely affecting creep properties. For this purpose 1% molybdenum was selected. Finally, sufficient silicon was added to further increase strength and creep resistance mainly by dissolving in α in which there is evidence that it segregates to, and reduces the mobility of, dislocations.

A parallel development in the United States led to the alloy Ti-8Al-1Mo-1V (Ti 8-1-1) which had a lower density, was weldable and had better forging characteristics because of the higher content of β-stabilizing elements. However, this alloy has an ordering parameter in excess of 9 which has led to problems of instability and loss of ductility due to the tendency to form the α_2-phase after long time exposure at elevated temperatures. More recently another American alloy has been developed which offers a compromise between the above two alloys. This is commonly known as Ti-6242 and has the composition Ti-6Al-2Sn-4Zr-2Mo-0.1Si.

All these alloys are forged at temperatures that place them well within the $\alpha + \beta$ phase field. The recommended heat treatment is to solution treat at a temperature at which the alloy consists of approximately equal proportions of α- and β-phases, e.g. 900°C for IMI 679 and 1010°C for Ti 8-1-1. For maximum creep strength, the alloys are then air-cooled to form a microstructure of equi-axed grains of primary α and Widmanstätten α which forms by nucleation and growth from β (Fig. 6.6a). Faster cooling will cause the high temperature β-phase to transform to martensitic α', at least in thin sections, which causes some increase in tensile strength although creep resistance is reduced at the upper end of the temperature range ($>450°$C). The alloys are then normally given a stabilizing heat treatment within the range 500 to 590°C.

β-heat treated alloys   Forging of titanium alloys in the β-phase field offers the advantage of easier deformation because of the higher working temperatures and the fact that the alloys have a body-centred cubic

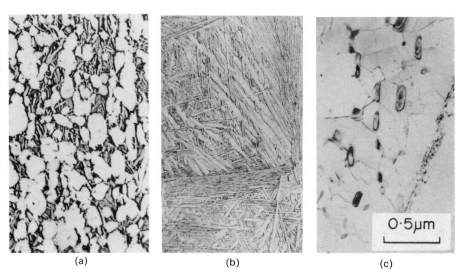

(a)   (b)   (c)

Fig. 6.6a Alloy IMI 679 air-cooled from the $\alpha + \beta$ phase field. The white phase is
primary α and the other is Widmanstätten α. ×500; b, alloy IMI 685 oil-quenched
from the β-phase field showing laths of the martensitic α'-phase delineated by small
amounts of the β-phase. ×75; c, IMI 685 quenched from the β-phase field and aged at
850° C showing particles of the phase $(TiZr)_5Si_3$. ×30 000 (courtesy IMI Titanium)

structure. However this practice, and the subsequent heat treatment of the
alloys in this phase field, is normally avoided because excessive grain
growth may occur which adversely affects ductility at room temperature.

The near-α alloy IMI 685 (Ti-6Al-5Zr-0.5Mo-0.25Si) is an example of a
composition developed to explore the opportunities of both β-forging and
β-heat treatment. The α/β transus is 1020° C and quenching from 1050° C
produces laths of martensitic α' which are delineated by thin films of β that
are retained (Fig. 6.6b). Subsequent ageing at 500–550° C reduces quenching stresses and causes some strengthening. Martensitic α' transforms to α
and the microstructure comprises laths of α bounded by a fine dispersion of
particles. Electron diffraction studies have indicated that these particles
may be either body-centred cubic ($a = 0.33$ nm) which is the normal β-
titanium structure, or face-centred cubic ($a = 0.44$ nm). The β-particles are
considered to form by spheroidization of the inter-lath films but little is
known of the other phase except that it contains titanium, molybdenum
and silicon. If ageing is carried out at higher temperatures, e.g. 850° C,
softening occurs and it is possible to observe the precipitate $(Ti, Zr)_5Si_3$
which forms on dislocation networks in the boundaries between the α-laths
(Fig. 6.6c).

Creep resistance is high in the range 450 to 520° C and is at a maximum if
intermediate quenching rates in the range 1 to 10° C s^{-1} are used. In this
condition the basket weave morphology, e.g. Fig. 6.4d, is present in the
microstructure. In thick sections, or with slow quenching rates, the

microstructure can contain coarse, aligned laths of α which reduce room temperature ductility and increase the rate of crack propagation in low-cycle fatigue (Section 6.7.2).

As compared with earlier alloys, the essential features of the composition IMI 685 are as follows:

(i) Tin is replaced by a lower amount of aluminium to reduce density whilst maintaining the ordering parameter within safe limits.

(ii) The content of the β-stabilizing element molybdenum is halved which reduces the amount of the β-phase, the presence of which leads to lower creep resistance.

(iii) Zirconium is added to provide solid solution strengthening of α.

(iv) The level of silicon is increased slightly to allow for its greater solubility at the higher β-heat treatment temperature.

More recently, the near-α alloy IMI 829 has been developed with enhanced creep strength at temperatures up to 600° C (Fig. 6.2). This alloy is also heat treated in the β-phase field and has the nominal composition Ti-5.5Al-3.5Sn-3Zr-0.25Mo-1Nb-0.3Si. It also has superior oxidation resistance to the earlier near-α alloys which is important as this problem is likely to be a limiting factor on long-time exposure at such temperatures. IMI 829 is being evaluated for gas turbine components.

6.2.4 Ti-Cu age-hardening alloy

Although Ti-Cu has no commercial significance as a β-eutectoid system, it was recognized that the titanium-rich end of the phase diagram offered potential for developing an alloy that may respond to age hardening. This follows because the solubility of copper in α-titanium reduces from 2.1 % at the eutectoid temperature of 798° C to 0.7 % at 600° C and to a very low value at room temperature. Moreover, it seemed possible that such an alloy could be cold-formed after solution treatment when in a relatively soft condition and then strengthened by ageing.

Ti-Cu alloys were investigated in Britain where hot-forming facilities were limited and the composition Ti-2.5Cu (IMI 230) was developed as a heat treatable sheet material. It is of special interest as it is one of very few titanium alloys that is strengthened by a classical age-hardening reaction. Solution treatment is carried out at 805° C and is followed by air-cooling (sheet) or oil-quenching to room temperature. A double ageing treatment at 400 and 475° C promotes precipitation of a fine dispersion of the metastable form of the phase Ti_2Cu which is coherent with the α-matrix (Fig. 6.7) and forms on the $\{10\bar{1}1\}$ planes.

A moderate increase of 150–170 MPa in tensile strength may be achieved by ageing. Strength properties are further enhanced if the alloys are cold-formed prior to ageing and, in this condition, compare favourably with the α-alloy Ti-5Al-2.5Sn which requires hot-forming. Ti-2.5Cu is weldable and strength may be recovered providing the duplex ageing treatment is applied after welding. Fig. 6.8 shows an application which is a casing for a gas turbine engine that is constructed by welding together forged rings and vanes formed from IMI 230 sheet.

Fig. 6.7 Coherent plates of Ti_2Cu zones in an aged Ti-2.5Cu (IMI 230 alloy) (courtesy IMI Titanium)

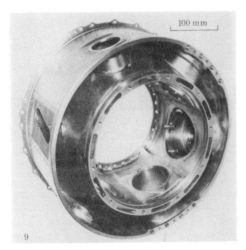

Fig. 6.8 Welded IMI 230 alloy casing from a Rolls Royce/SNECMA Olympus engine (courtesy Rolls Royce Ltd)

It will be noted that the commercial composition is a binary alloy and it might be anticipated that additions of other elements, singly or together, may induce a further response to age hardening. However, extensive investigation has failed to reveal any addition having this desired effect.

6.3 α/β alloys

The limitations in strength that can be developed in the fully-α alloys because of the ordering reaction occurring at higher solute contents, together with difficulties with hot-forming, led to the early investigation of compositions containing both the α- and β-phases. These α/β alloys now have the greatest commercial importance with one composition, Ti-6Al-4V

Fig. 6.9 Forged IMI 318 (Ti-6Al-4V) blades from the LP rotor stage of the Rolls Royce/SNECMA Olympus 593 jet engine (courtesy Rolls Royce Ltd)

(IMI 318), making up more than half the sales of titanium alloys both in Europe and the United States. They offer the prospect of relatively high tensile strengths and improved formability, although some sacrifice in creep strength occurs above 400°C as well as reduced weldability. Their principal use is for forged components, e.g. in the fan blades of jet engines (Fig. 6.9).

Most α/β alloys contain elements to stabilize and strengthen the α-phase, together with 4 to 6% of β-stabilizing elements which allow substantial amounts of this phase to be retained on quenching from the β or $\alpha + \beta$ phase fields. The common β-stabilizing elements confer solid solution strengthening of the β-phase although, as shown in Table 6.2, these effects are relatively small. This table also gives the minimum solute concentration needed to give complete retention of metastable β on quenching of binary alloys to room temperature. Strength properties of α/β alloys may be enhanced by subsequent tempering or ageing treatments and room temperature tensile strengths exceeding 1400 MPa have been achieved. However, few compositions can sustain these levels of strength in thick sections because of hardenability effects on quenching which are aggravated by the low thermal conductivity of titanium (Table 1.1).

It is proposed now to consider structure/property relationships in α/β alloys which are developed by heat treatment. Reference should again be made to Fig. 6.3 although some additional phase transformations will need to be considered.

Table 6.2 Solid solution strengthening and β-stabilizing capacity of β-stabilizing alloying elements (from Hammond, C. and Nutting, J., *Metal Science*, **11**, 474, 1977)

	Element							
	V	Cr	Mn	Fe	Co	Ni	Cu	Mo
Solid solution strengthening (MPa wt%$^{-1}$)	19	21	34	46	48	35	14	27
Minimum alloy content to retain β on quenching (%)	14.9	6.3	6.4	3.5	7	9	13	10

6.3.1 Annealed alloys

For alloys which have phase diagrams of the β-isomorphous type, uniform properties can be obtained in thick sections by slow cooling from either the β or $\alpha + \beta$ phase fields, known as the β-annealed and mill-annealed conditions respectively. In the first case, it is usual for the α-phase to form as Widmanstätten laths in a β-matrix, although β may itself transform to martensitic α'. The size of the laths depends on the rate of cooling and the basket weave structure is again obtained when cooling rates are slow (Fig. 6.10a). Annealing in the $\alpha + \beta$ phase field is usually carried out at about 700°C and, in addition to providing stress-relief, this treatment results in the formation of an equi-axed structure composed of α-grains and grains of transformed β (Fig. 6.10b). These latter grains transform to Widmanstätten α as is evident in the transmission electron micrograph shown in Fig. 6.10c. Grain size can be modified by suitable adjustment of working and annealing cycles and the amount of primary α is dictated by the lever rule. Frequently a second, or duplex anneal, is given which causes further partitioning of alloying elements between the α- and β-phases, and the main purpose of this treatment is to enrich the β-phase which increases the stability of the alloys for service at elevated temperatures.

The α/β titanium alloys are most often used in the annealed condition and it should be noted that both microstructure and some mechanical properties may differ depending upon whether or not prior forming was carried out above or below the β-transus. Table 6.3 compares the properties of the alloy Ti-6Al-4V forged in these two conditions. It should be noted that, although tensile properties are fairly similar, the samples forged in the $\alpha + \beta$ phase field (equi-axed grains) are more ductile, whereas fracture toughness and fatigue strength are both notably higher in β-forged and annealed material (acicular Widmanstätten structure). Work on Ti-6Al-4V rolled plate has indicated that the superior fatigue performance with the β-annealed condition is associated with relatively slower rates of crack propagation (Fig. 6.11a). This effect, in turn, is attributed to the slower progress of cracks through the Widmanstätten microstructure, particularly at stress intensities below a critical value (T in Fig. 6.11a) at which desirable crack branching occurs within packets of the α-laths (Fig. 6.11b). These

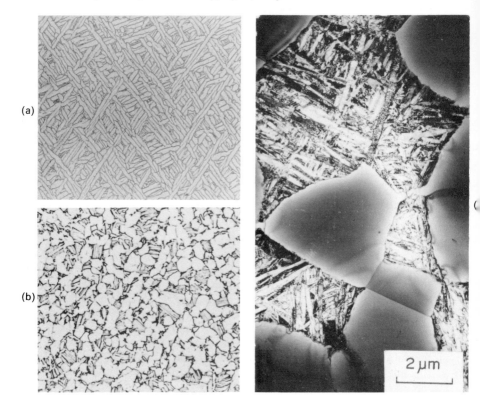

Fig. 6.10a Alloy IMI 318 (Ti-6Al-4V) slowly cooled from β-phase field showing basket weave structure of Widmanstätten α-plates in a β-matrix (courtesy Rolls Royce Ltd). ×320; b, alloy IMI 318 annealed at 700°C in α +β phase field. Equi-axed grains of α (white) and transformed β (Widmanstätten α) (courtesy IMI Titanium). ×500; c, transmission electron micrograph of (b) showing the structure of the transformed β (Widmanstätten α) (courtesy C. Hammond). ×7500

trends have already been noted when considering α-alloys and appear to apply generally in the α/β alloys.

In the β-eutectoid alloys, very slow cooling from the β-phase field leads to the formation of a lamellar eutectoid of α and a compound such as Ti_2Cu, in a manner that is analogous to pearlite formation in steels. However, these structures have so far found no application in commercial titanium alloys because the reactions are sluggish and the phases that form cause embrittlement.

6.3.2 Quenching from β-phase field

The range of properties of α/β alloys can be extended by quenching from

Table 6.3 Properties of annealed Ti-6Al-4V forgings●

	Forging treatment ■	
	$\alpha + \beta$ phase field	β phase field
Tensile ultimate (MPa)	978	991
Tensile yield (MPa)	940	912
Tensile elongation (%)	16	12
Reduction in area (%)	45	22
Fracture toughness (MPa m$^{1/2}$)	52	79
10^7 fatigue limit (MPa)▲	± 494	± 744

● Annealed 2 hours at 705°C, air-cooled after forging
■ α/β transus 1005°C
▲ Axial loading: smooth specimens, $K_t = 1.0$

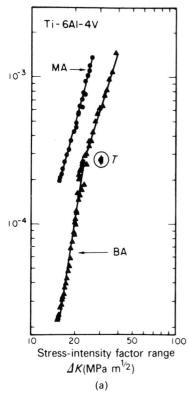

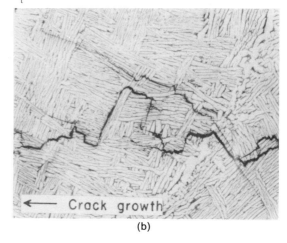

Ti-6Al-4V

MA

T

BA

Stress-intensity factor range
ΔK(MPa m$^{1/2}$)

(a)

Crack growth

(b)

Fig. 6.11a Fatigue crack growth rates for Ti-6Al-4V rolled plates in the β-annealed (BA) and mill-annealed (MA) conditions. BA = 0.5 h 1038°C, air-cool to room temperature. Tests conducted at 5 Hz using compact tension specimens. Ratio of minimum to maximum load = 0.1 (from Yoder, G. R. *et al.*, *Met. Trans.*, **8A**, 1937, 1977); b, branching of fatigue cracks within the Widmanstätten packets of the α-laths (courtesy J. Ruppen and A. J. McEvily). ×225

the β-phase field and then tempering or ageing at elevated temperatures to decompose the quenched structures. The changes that occur may be complex and it is necessary to study them in some detail.

A distinction can be made between relatively dilute α/β alloys which form hexagonal α' martensite or two orthorhombic martensites α'' and α''' on quenching, and more concentrated alloys in which the β-phase may be partly or completely retained in a metastable condition. The division between the two types of behaviour can be shown by the M_s (martensite start) line that is included in Fig. 6.1b,c and Fig. 6.3. If the alloys contain a sufficient content of β-stabilizing elements to bring the M_s temperature below room temperature then a fully metastable β-structure can be retained. The possible reactions (with the most important being underlined) and microstructures may be summarized as follows:

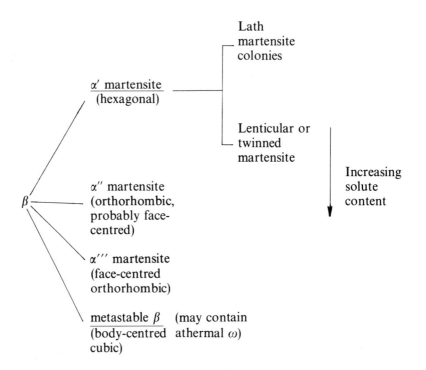

The most common martensite is the hexagonal α' type. In the more dilute alloys, it forms as colonies of parallel-sided plates or laths (Fig. 6.12a), the boundaries of which consist of walls of dislocations. The internal regions are also heavily dislocated. The structure is similar to that obtained when fully-α alloys are quenched from the β-phase field except that, in the α/β alloys, the laths are separated by thin layers of retained β-phase which is enriched in β-stabilizing solute elements. With increasing solute content and lower M_s temperatures, these colonies decrease in size and may degenerate into individual plates which are randomly oriented. These plates have a lenticular or acicular morphology (Fig. 6.12b) and are internally twinned on $\{10\bar{1}1\}_{\alpha'}$ planes. The orientation relationship of the β-phase and α' martensite is $(110)_\beta//(0001)_{\alpha'};\ [111]_\beta//[11\bar{2}0]_{\alpha'}$, and the habit planes for the untwinned and twinned planes are $\{334\}_\beta$ and $\{344\}_\beta$ respectively.

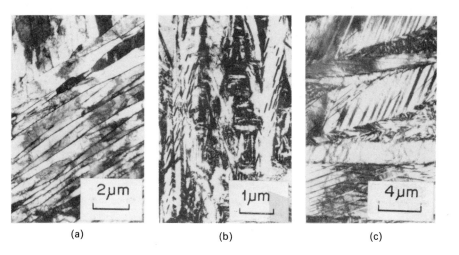

(a)   (b)   (c)

Fig. 6.12 Transmission electron micrographs showing the structure of titanium alloy martensites: a, hexagonal α′ (lath) martensite in a dilute alloy (Ti-1.8Cu) quenched from 900°C; b, hexagonal α′ (lenticular) martensite containing twins in a concentrated alloy (Ti-12V) quenched from 900°C; c, orthorhombic α″ martensite in the alloy Ti-8.5Mo-0.5Si quenched from 950°C (from Williams, J. C., In: *Titanium Science and Technology*, ed. R. I. Jaffee and H. M. Burte, Plenum Press, New York, Volume 3, 1973)

The second type of titanium martensite (α″) has an orthorhombic structure and a similar lattice correspondence with the β-phase. For this reason it is probably face-centred orthorhombic and its lattice dimensions are $a = 0.298$ nm, $b = 0.494$ nm, $c = 0.464$ nm. It is also internally twinned (Fig. 6.12c), the twins forming on $\{111\}_{\alpha''}$ planes. Formation of α″ is thought to be strongly composition dependent, e.g. it occurs in Ti-Mo but not in Ti-V alloys, although it may be stabilized in the latter alloys by adding aluminium as in some Ti-V-Al alloys. α″ has also been observed in certain other highly alloyed α/β compositions but it is unlikely that its tempering behaviour will assume commercial importance in the future.

Two other martensites (face-centred orthorhombic and face-centred cubic) have been reported in electron microscope studies and the former has been confirmed by X-ray diffraction techniques. It has been termed α‴ (or β′) and has lattice parameters quite distinct from α″ martensite, i.e. $a = 0.356$ nm, $b = 0.439$ nm and $c = 0.447$ nm. The orientation relationship with the β-phase is $(0\bar{1}1)_{\beta}//(001)_{\alpha'''}$; $[\bar{1}11]_{\beta}//[\bar{1}10]_{\alpha'''}$ and the habit plane is close to the $\{133\}_{\beta}$ planes. The existence of the second phase has not been confirmed and it may well arise as an artefact due to the use of thin metal foils for examination in the electron microscope.

If the M_s and M_f fall above and below room temperature respectively, then a mixed microstructure containing lenticular α' or α'' (or perhaps α''') martensite may be formed together with retained β. Another feature is that metastable β may contain a fine dispersion of a phase ω, the formation of which cannot be suppressed even at fast quenching rates. The nature of this athermal ω is discussed in Section 6.3.4.

As mentioned earlier, lack of hardenability can cause variations in the strength and microstructure of thick sections that are quenched and, in such cases, solution treating within the $\alpha + \beta$ rather than the β-phase field can be an advantage. Such treatments cause partitioning of solute between α and, more particularly, the reduced volumes of β, thereby increasing the stability of this latter phase so that it is less likely to transform during quenching. This effect increases as the solution treatment temperature is lowered because the volume of β-phase is further reduced, an observation that has led to the practice known as soft quenching, which has been applied to α/β alloys of large section (up to 150 mm). This practice involves solution treating at a temperature high in the $\alpha + \beta$ field to dissolve the alloying elements and slow cooling to $700°C$, at a rate of $50-150°C$ an hour, to allow partitioning of solute as the amount of β-phase is reduced. It is then possible to air-cool the alloy to room temperature and still retain the β-phase.

Quenched alloys are normally tempered or aged to decompose the retained phases and it is now appropriate to consider the changes that may occur.

6.3.3 Tempering[†] of titanium martensites

Titanium martensites transform on heating at elevated temperatures by several reactions the nature of which depends upon the crystal structure of the martensite and the composition of the alloy concerned. The reactions may be complex and the various types are shown below.

These reactions lead to a wide range of microstructures. In the β-isomorphous alloys, α' decomposes directly to α of equilibrium composition at the tempering temperature and β forms as a fine precipitate that is nucleated heterogeneously at martensite plate boundaries or at internal substructures such as twins. Significant increases in strength may result (Table 6.1). In β-eutectoid alloys, α' may decompose directly into the α-phase and an intermetallic compound, although the formation of this compound may take place in several stages. However, in systems such as Ti-Mn where the normal eutectoid reaction is sluggish, the martensite tempers first by forming α and precipitates of β with the intermetallic compound appearing only slowly at a later stage.

Tempering of α'' martensite may occur by two mechanisms. In alloy

[†] Quenched alloys are normally heated for a period of time at an elevated temperature. This treatment is referred to as tempering in the case of martensites, and ageing when retained β is being decomposed, although the actual heat treatments are similar.

α' martensite

β-isomorphous alloys $\alpha' \rightarrow \alpha + \beta$

β-eutectoid alloys

$$\begin{cases} \text{Alloys with} \\ \text{slow eutectoid} \quad \alpha' \rightarrow \alpha + \beta \rightarrow \alpha + \text{compound} \\ \text{reactions} \\ \text{e.g. Ti-Mn} \\[2ex] \text{Alloys with} \\ \text{fast eutectoid} \quad \sigma' \rightarrow \alpha + \text{compound} \\ \text{reactions} \qquad\qquad\quad \text{(may form in} \\ \text{e.g. Ti-Cu} \qquad\qquad\quad \text{several stages)} \end{cases}$$

α″ martensite

$$\begin{cases} \text{Alloys with high} \\ M_s(\alpha'') \text{ temperature} \\ \alpha'' \rightarrow \alpha'' + \alpha \rightarrow \alpha'' + \alpha + (\alpha + \beta) \rightarrow \alpha + \beta \\ \qquad\qquad\qquad\qquad\quad \text{(cellular} \\ \qquad\qquad\qquad\qquad\quad \text{reaction)} \\[2ex] \text{Alloys with low} \\ M_s(\alpha'') \text{ temperature} \quad \alpha'' \rightarrow \beta \rightarrow \text{products} \\ \qquad\qquad\qquad\qquad \text{(see Section} \\ \qquad\qquad\qquad\qquad 6.3.4) \end{cases}$$

compositions in which M_s (α'') occurs at relatively high temperature, α'' decomposes first by the formation of a fine and uniformly dispersed α-phase in the α''-matrix. Further ageing causes both the coarsening of these particles and the nucleation of a cellular reaction at the prior β-grain boundaries leading to the growth of a lamellar structure of $\alpha + \beta$. Growth of these lamellar cells then occurs at the expense of the other regions. In alloys having an M_s (α'') temperature near to room temperature, the α'' reverts to the β-phase which then decomposes by a mechanism which is characteristic of the particular tempering temperature. Decomposition of the β-phase is discussed in Section 6.3.4.

No information is available concerning the tempering of α''' martensite but its similar crystal structure suggests it may behave like α''.

6.3.4 Decomposition of metastable β

Decomposition of the β-phase that is retained on quenching occurs on ageing at elevated temperatures. It is frequently the dominant factor in the heat treatment of α/β and β-alloys, particularly when the aim is to develop high tensile strength (Fig. 6.3). The direct transformation of β to the equilibrium α-phase occurs only at relatively high temperatures probably because of the difficulty of nucleating the close-packed hexagonal α-phase from body-centred cubic β-matrix. Accordingly, intermediate decomposition products are usually formed and the possible reactions are summarized and discussed below.

Medium alloy content

   100–500°C $\beta \to \beta + \omega \to \beta + \alpha$

Concentrated alloys

   200–500°C $\beta \to \beta + \beta_1 \to \beta + \alpha$
   > 500°C $\beta \to \beta + \alpha$

ω-phase   As mentioned earlier, athermal ω may form in the β-phase during quenching of some compositions and it occurs by a displacement reaction. More commonly, however, ω precipitates isothermally as a very fine dispersion of particles when alloys containing metastable β are isothermally aged at temperatures in the range 100–500°C. The ranges of stability of both types of ω are shown schematically for a β-isomorphous phase diagram in Fig. 6.13, but there is some evidence that athermal ω can also form during heating to the isothermal ageing temperature.

The ω-phase has attracted special attention because its presence can cause severe embrittlement of the alloy concerned. In a more positive vein, ω particles have been beneficial in the specialized field of superconducting titanium alloys as they are effective in flux-pinning, with a consequent large improvement in the critical current densities that may be sustained in the presence of an external magnetic field.

Studies of the isothermal ω-phase have revealed the following characteristics:

(i) ω forms rapidly as homogeneously nucleated, coherent precipitates with particle densities that may exceed 80% by volume (Fig. 6.14).

(ii) The ω-particles are cuboidal in shape if there is a high misfit with the β-matrix and ellipsoidal if the misfit is low.

(iii) Partitioning of solute occurs during ageing leading to depletion of ω and enrichment of the β-matrix. The terminal composition of ω in aged binary titanium alloys is related to the group number of the solute in the Periodic Table because the electron:atom ratios of all ω-phases have been found close to 4.2:1. Thus the possibility exists that ω is an electron compound.

(iv) Most results suggest that ω has an hexagonal structure with a constant c/a ratio of 0.613 for all systems in which it is found.

(v) Dislocations have little or no mobility in ω which accounts for the embrittlement of alloys having high volume fractions of this phase. It is interesting to note however that, even in alloys displaying no macroscopic ductility, the fracture surfaces show exceedingly small dimples which are indicative of some ductility at a microscopic scale. Thus the possibility exists that the potent hardening associated with ω may be used to practical advantage although no progress has been made in this regard.

The formation of isothermal ω may be minimized or avoided by control of the ageing conditions, as well as by varying alloy composition. The significance of both ageing temperature and composition is apparent from Fig. 6.13. The upper temperature limit of stability of ω in most binary alloys is close to 475°C and the range of stability decreases as the solute content is raised. This latter effect is attributed to a relative increase in the stability of the β-phase and it should be noted that this effect can arise from the presence of both α- and β-stabilizing elements. For example, ω is formed in binary Ti-V alloys, but is absent in the important ternary alloy Ti-6Al-4V. This is one reason why most α/β and β-titanium alloys contain at least 3% aluminium.

β-phase separation Separation of the β-phase into two bcc phases of different compositions is favoured in alloys which contain sufficient

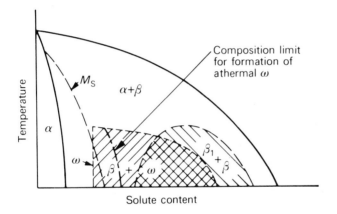

Fig. 6.13 Schematic β-isomorphous alloy phase diagram showing an M_s curve and the ranges of stability of ω. β and β_1

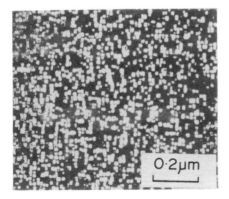

Fig. 6.14 Dense dispersion of cuboids of the ω-phase in a Ti-11.5Mo-4.5Sn-6Zr alloy (from Williams, J. C., In: *Titanium Science and Technology*, ed. R. I. Jaffee and H. M. Burte, Plenum Press, New York, Volume 3, 1973)

β-stabilizer to prevent ω formation during low temperature ageing, and which transform only slowly to the equilibrium phase α under these conditions (Fig. 6.13). This transformation is thought to occur during ageing of a wide range of alloys but has received much less attention than the $\beta \rightarrow \omega$ reaction because it is not considered to be important in commercial alloys. The phases that form have been designated as β(matrix) and $\beta_1{}^\dagger$, which occurs as a uniformly dispersed, coherent precipitate. Again there is partitioning of solute between the two phases which leads to enrichment of the β-matrix and depletion of β_1 during the ageing treatment.

Formation of equilibrium α-phase   Ageing of alloys containing metastable β can, under certain circumstances, result in the direct nucleation of the α-phase. Alternatively this phase may form indirectly from either the ω- or β_1-phases. The route that is followed controls morphology and distribution of α and thus has a marked effect on properties. α that forms directly from β can have two distinct morphologies. It may occur as coarse Widmanstätten plates in a β-matrix in both the relatively dilute binary alloys aged at temperatures above the range for ω formation, and in more complex alloys containing substantial amounts of aluminium, e.g. Fig. 6.10b,c. In such cases, ductility may be adversely affected and deformation prior to ageing is desirable so as to obtain a more uniform distribution of α. Alternatively, a fine dispersion of α in a β-matrix is obtained when alloys containing high concentrations of β-stabilizing elements are aged at temperatures above which phase separation of β occurs.

When α forms in alloys comprising $\beta + \omega$ microstructure, the mechanism of nucleation depends upon both the relative misfit between these two phases as well as the ageing temperature. If the misfit is low, α nucleates with difficulty and it forms by a cellular reaction that occurs heterogeneously at the β-grain boundaries. If the misfit is high then α nucleates at the β/ω interfaces. High ageing temperatures encourage α to form directly from ω.

Continued ageing of alloys that have undergone β-phase separation into $\beta + \beta_1$ leads to the nucleation of the α-phase within the β-particles. Thus the final α-phase distribution is determined by the distribution of β_1, and so is characteristically uniform and closely spaced. Although this reaction may, inadvertently, play a part in the heat treatment of a number of commercial α/β titanium alloys, it has not been studied in detail.

6.3.5 Fully heat-treated α/β alloys

As mentioned earlier, the most commonly used titanium alloy is Ti-6Al-4V which has many applications as a general purpose alloy and is usually used in the annealed condition. For applications such as fasteners that require higher strength, the alloy is solution treated high in the $\alpha + \beta$ phase field, quenched, and aged or annealed at around $700°$C. The resultant structure

---

† Occasionally these phases are designated as β_1 and β_2, respectively.

then comprises equi-axed α-grains in a matrix of fully transformed β (Fig. 6.10b,c). The alloy combines a minimum tensile strength of about 960 MPa with good creep resistance up to 380°C. It is particularly used in forgings but is also available in plate, sheet, rod or wire forms.

For applications such as aircraft engine mounting brackets and undercarriage components greater strength is required and a number of other α/β forging alloys have been developed. More use has been made of the heat treatment potential of the alloys and, in general, the content of stabilizing elements has been increased so that the M_f temperature is depressed well below room temperature and significant amounts of β are retained, e.g. the alloys Ti-6Al-6V-2Zr-0.7 (Fe, Cu) (Ti-662), Ti-6Al-2Sn-4Zr-6Mo (Ti-6246), Ti-4Al-4Sn-4Mo-0.5Si (IMI 551) and Corona-5 (Ti-4.5Al-5Mo-1.5Cr). Full strengthening is again usually obtained by solution treating in the $\alpha + \beta$ phase field, quenching and ageing to transform the retained β. On the other hand, fracture toughness is higher if the alloys are first cooled from the β-phase field (see Table 6.2). For example, the newly developed alloy Corona-5 has a plane strain fracture toughness as high as 155 MPa m$^{\frac{1}{2}}$, for a tensile strength of 950 MPa if it is first annealed in the β-phase field. This value for fracture toughness is double that normally obtained with Ti-6Al-4V and is again attributed to the slower rate of crack propagation, through the microstructure containing elongated Widmanstätten laths of the α-phase, e.g. Fig. 6.10a and Fig. 6.11b.

6.4 β-alloys

The addition of sufficient β-stabilizing elements to titanium can produce a fully β-structure at room temperature (see Fig. 6.1 and Table 6.2) and alloys of this type attracted early attention because of the superior forming characteristics anticipated for the body-centred cubic structure. In particular, they offered the prospect of being cold-formed in a relatively soft condition and then strengthened by age hardening. Another potential advantage was that the presence of a high content of solute elements increased hardenability which may allow the through hardening of thick sections during heat treatment. Such high contents of β-stabilizing elements did, however, cause problems with ingot segregation and tended to increase density, some alloys having relative densities in excess of 5.

The first alloy to be used commercially was the composition Ti-13V-11Cr-3Al which may be solution treated, quenched, cold-formed and aged at 480°C to give a tensile strength as high as 1300 MPa. Strengthening is due mainly to a combination of solid solution hardening of the β-phase and age hardening, the latter arising from precipitation of a fine dispersion of the α-phase in the β-matrix. This alloy has limited weldability due mainly to formation of the ω-phase in the heat-affected zones. Moreover, it has proved to be unstable if exposed for long times above 200°C as such a treatment causes precipitation of the compound $TiCr_2$ which reduces the toughness of the alloy. It should be noted, however, that the above alloy was successfully used for the skin and some structural members of the American

SR-71 aircraft designed to fly at speeds around 3200 km h^{-1} at which aerodynamic heating precludes the use of aluminium alloys. Another relatively early β-alloy was the medium-strength British composition Ti-15Mo (IMI 205) which has found application in the chemical industry.

More recently, other β-alloys have been developed for which the formulation of compositions has been controlled mainly by these guidelines:

(i) The addition of elements, e.g. aluminium, zirconium and tin, which tend to suppress or restrict the formation of the ω-phase during heat treatment or welding by promoting nucleation of the α-phase.

(ii) Limiting the amount of elements, e.g. chromium, which tend to stabilize the β-eutectoid transformation and cause embrittlement because of the formation of $TiCr_2$ or other compounds. It should also be noted that excessive eutectoid stabilization leads to a sluggish response to age hardening which is undesirable.

(iii) Promotion of plastic strain by slip rather than by a mechanism involving a strain-induced transformation to martensite. Alloys which deform primarily by slip have been found to possess better forming characteristics.

Examples of more recent alloys are the American compositions Ti-8Mo-8V-2Fe-3Al (Ti 8823), Ti-15V-3Sn-3Cr-3Al (Ti 15-3-3-3) and Ti-11.5Mo-6Zr-4.5Sn (Beta III). A high level of strengthening is possible in each alloy which again arises primarily from a combination of solid solution hardening and age hardening. Some cold-deformation of alloys such as Beta III can be carried out at room temperature prior to ageing, which may further enhance strength properties due to the nucleation of finely dispersed α-precipitates on dislocations.

A typical heat treatment cycle for Beta III is solution treatment at 750 to 775° C, then water quench or air-cool to room temperature and age 24 h at 475° C to precipitate α in a β-matrix. The ageing temperature is above that at which ω should form in this alloy. Tensile strengths as high as 1400 MPa combined with a fracture toughness of 54 MPa $m^{\frac{1}{2}}$ have been obtained and the alloy has found particular use for high-strength fasteners.

Compared with the α and α/β titanium alloys, the β-alloys have so far found few applications. However, now most of the deficiencies of the early compositions have been overcome it is expected that the high strength, good formability and high hardenability of this class of alloys will result in greater use being made of them. In this regard, the relatively low flow stresses exhibited by the β-alloys Ti 15-3-3-3 and Ti 10-2-3 (Ti-10V-2Fe-3Al) during hot-working at a wide range of strain rates are apparent when they are compared with the α/β alloy Ti-6Al-4V in Fig. 6.15. The alloy Ti 10-2-3 has particularly good forging characteristics which suggests that it has potential for thick forgings for the aircraft industry. It was also observed that the flow stresses for all three alloys decrease to plateau values as the strain rates fall below approximately 10^{-5} s^{-1}. These values correspond to the onset of superplastic behaviour (Section 6.5.1).

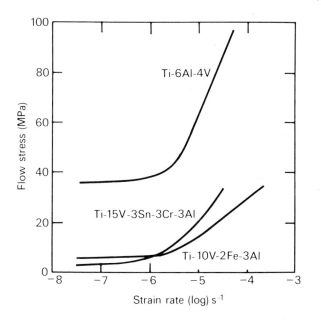

Fig. 6.15 Relationship between flow stress and strain rate for titanium alloys hot-worked at 810° C (courtesy H. W. Rosenberg)

6.5 Fabrication

6.5.1 Hot-working

The as-cast structure of consumable-arc melted ingots is sensitive to cracking and the initial working is normally done by hot-press forging. Deformation is carried out at a relatively slow rate in large hydraulic presses and the ingots may be press forged to slab about 150 mm thick for subsequent rolling to plate or sheet. Alternatively, they may be pressed to round or square billets for processing to bar, rod, tube, extruded sections or wire. Rough forgings for components such as gas turbine compressor discs can be pressed directly from ingots.

Above 550°C, titanium will absorb oxygen to form both an oxide scale and a brittle, sub-surface layer which can initiate surface cracks. Titanium also absorbs hydrogen and it is necessary to ensure that furnace atmospheres are hydrogen-free. For this reason, electric preheating furnaces are preferable to those fired by oil or gas. Chemical descaling, abrasive cleaning and even machining, frequently combined with careful surface inspection, may be necessary before any further working operations are carried out. Further descaling and cleaning may also be required after later stages of hot-working and heat treatment.

Titanium alloys can be hot-worked to produce most of the shapes that can be obtained with steels and other metals and they are often equated with stainless steels when comparing their hot-working characteristics. Power requirements are usually relatively high, particularly when microstructural considerations dictate that the working operations be conducted below the α/β transus temperature. Another factor is the narrow temperature ranges over which most hot-working operations must be carried out.

Although some pre-forming of titanium alloys may be undertaken by open-die forging, most forging operations normally involve closed dies in which the shaping of hot metal occurs completely within the walls or cavities of the two dies as they come together. Dies for closed-die forging are normally preheated to 200–250°C for rapid operations involving hammers or mechanical presses, and to around 425°C for slower working in hydraulic presses.

Heating of dies is particularly critical when forming thin sections, e.g. sheet, otherwise the poor thermal conductivity of titanium can lead to localized chilling which causes uneven metal flow or even cracking in a workpiece. Some novel hot-forming techniques have been developed. For example, assemblies of quite large dimensions can be produced by slow, isothermal forming (or creep forming) between heated dies, some of which can be made cheaply from cast ceramics. Metallic tooling is much more expensive, particularly as the die faces may need to be made from nickel or cobalt alloys in order that hardness and oxidation resistance are adequate at the relatively high forming temperatures, e.g. 850°C.

If creep forming is carried out at temperatures close to 0.6 T_M, where T_M is the melting temperature of an alloy in degrees K, and at controlled strain rates of around 10^{-5} to $10^{-6} s^{-1}$, then alloys having stable, small grain sizes may exhibit superplasticity. Flow stresses become very low (see Fig. 6.15) and components can be produced by simple methods similar to those used for thermoforming plastics. Typical process parameters for the α/β alloy Ti-6Al-4V are 100 to 1000 KPa pressure at 900 to 950°C, applied for times of 0.25 to 4 h. Such pressures, which should be even lower for β-alloys, can readily be achieved using an inert gas, e.g. argon, that may be introduced into one part of the mould cavity, with the titanium sheet serving as a deformable diaphragm that flows into the other part.

Clean titanium can readily diffusion bond to itself under conditions very similar to those used for superplastic forming (Section 6.5.5). Thus forming and joining can be combined in the one comparatively simple operation to produce special products, e.g. the truss section shown diagrammatically in Fig. 6.16.

Titanium alloys are particularly susceptible to galling, i.e. wear due to friction, during hot- or cold-working, which causes surface damage. Lubrication is thus essential and care must be taken in selecting materials that do not react with titanium when heated. Suspensions of graphite or molybdenum disulphide are suitable for both types of operation, whereas glass may be required for more severe processes such as hot extrusion.

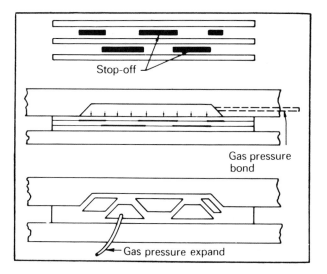

Stop-off

Gas pressure bond

Gas pressure expand

Fig. 6.16 Advanced manufacturing techniques using superplastic forming combined with diffusion bonding to produce a truss section. The stop-off material has been inserted where joining is not wanted (from Tupper, N. G. *et al., J. Metals*, **30**, No. 9, 7, 1978)

6.5.2 Cold-working

CP titanium and most alloys in the annealed condition have a limited capacity to be cold-worked. For example, minimum bend radii for sheet are commonly 1 to 3 times the gauge thickness (T) for CP titanium, 2 to 4 T for β-alloys and 3 to 6 T for most other alloys. One major problem is excessive springback which is a consequence of the low moduli and relatively high flow stresses of titanium and its alloys. To improve dimensional accuracy, cold-forming is generally followed by hot-sizing and stress-relieving for periods of 0.25 h at temperatures around 650–700°C. Such treatments may also help to restore strength properties that are reduced in certain directions of some deformed alloys containing the hexagonal α-phase in the manner described for wrought magnesium alloys in Section 5.6.1. The treatments may also cause changes to fine scale features of the microstructure, but little is known of these effects.

6.5.3 Texture effects

The major cause of property anisotropy in aluminium alloys, that of aligned, coarse intermetallic compounds (Fig. 2.18), is normally absent in titanium alloys. However, those titanium alloys which contain substantial amounts of the hexagonal α-phase may show marked elastic and plastic anisotropy if fabrication procedures produce a preferred orientation (or texture) in the grain structure. Such anisotropy will also be reflected in the mechanical properties and there is considerable interest in the prospect of controlled texture strengthening of titanium alloys. Such a process introduces a third dimension to alloy development as an adjunct to composition and microstructure.

Three easy slip modes and six twinning modes exist in α-titanium and the former are shown in Fig. 6.17a. The slip vectors in each case are parallel to the basal planes (0001) so that, if stress is applied in the direction of the *c*-axis, there will be no critical resolved shear stress acting in this plane. There are, however, other slip systems which can operate so that the requirement of having five independent slip systems for general plasticity is fulfilled. The elastic moduli in the *c*- and *a*-directions of single crystals of α-titanium show a large variation from 145 GPa to 99.5 GPa respectively. Although this difference is less in polycrystalline alloys showing preferred orientation, the elastic moduli in the longitudinal, long transverse and short transverse directions (Fig. 2.17) can still differ by as much as 30 %. An example of the variation in tensile properties with stressing direction is shown for the alloy Ti-6Al-4V in Table 6.4. In this case, maximum texture strengthening has occurred in the long transverse direction and is associated with the alignment of basal planes normal to the forging plane (Fig. 6.17b), which is one of the two favoured textures that may develop in rolled or forged titanium alloys. Fracture toughness and elastic modulus are also maximal in the long transverse direction.

Table 6.4 and Fig. 6.18 show that fatigue properties are lowest in the long transverse direction. This result has been attributed to the fact that Poisson's ratios are also sensitive to crystal orientation, these ratios being

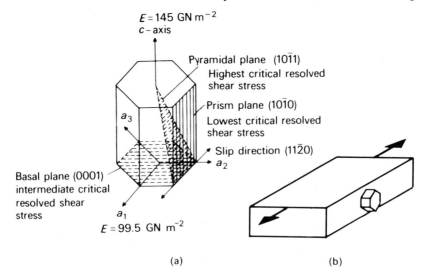

(a)   (b)

Fig. 6.17a Slip planes in α-titanium; b, alignment of hexagonal unit cell in α-titanium showing strongly preferred orientation after rolling

Table 6.4 Mechanical properties of 57 mm thick × 235 mm wide forged and annealed Ti-6Al-4V bar (from Bowen, A. W., In: *Titanium Science and Technology*, Vol. 2. ed. R. I. Jaffee and H. M. Burte, Plenum Press, New York, 1973; p. 1271)

Testing directions	0.2% proof stress (MPa)	Tensile strength (MPa)	Elastic modulus (GPa)	Elongation (%)	Approximate fatigue strength at 10^7 cycles (±MPa)
Longitudinal	834	910	114	17.5	496
Long transverse	934	986	128	17.0	427
Short transverse	893	978	114	12.5	565

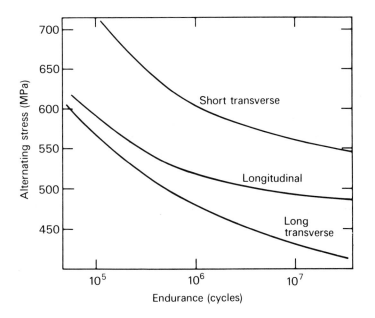

Fig. 6.18 Rotating-cantilever fatigue (*S/N*) curves for three testing directions in 57 mm thick, forged and annealed Ti-6Al-4V bar (from Bowen, A. W., In: *Titanium Science and Technology*, ed. R. I. Jaffee and H. M. Burte, Plenum Press, New York, Volume 2, 1973)

higher in the longitudinal and short transverse directions because stressing occurs parallel to the basal planes. Higher ratios imply greater constraint, which means that the levels of strain will be reduced and the fatigue strength enhanced in these two directions. The differences observed in fatigue strengths in the longitudinal and short transverse directions have been attributed to relative changes in grain shapes that also occur during processing.

The other type of texture that may be developed in rolled or forged titanium alloys involves alignment of basal planes parallel to the rolling or forging plane, i.e. the *c*-axis is parallel to the short transverse direction. This texture can be beneficial in sheet metal forming involving biaxial tension as thinning by simple slip becomes difficult. Measurements have shown that *R*-values (Section 3.3) are high and may lie in the range 1 to 5, whereas they are less than 1 for aluminium alloys. This form of texture strengthening is also useful in applications such as pressure vessels which require high biaxial strength.

6.5.4 Machining

Titanium alloys have unique machining characteristics. Whereas the cutting forces may be only slightly higher than those required for steels of equivalent hardness, there are other features that make these alloys relatively difficult to machine. In making comparisons, titanium alloys tend to be ranked together with austenitic stainless steels.

One basic problem arises because of the low thermal conductivity of titanium alloys which is only one-sixth that of steels. This feature, together with the fact that the characteristic shape of the chip allows only a small area of contact between chip and tool, means that high cutting temperatures may be generated during machining titanium alloys. For example, tests conducted at similar cutting speeds revealed that the temperature developed at the cutting edge of a tungsten carbide tool was 700°C for titanium alloys and only 540°C for a steel. Tool lives will thus be drastically shortened in the former case unless slower cutting speeds are used. Another feature associated with the machining of titanium alloys is the tendency for chips to stick to the cutting edge (galling), particularly once the tool becomes warm. This, in turn, may also reduce tool life as fracture of the cutting edge may occur if the titanium chip is removed when the tool re-enters the workpiece on the next pass. Alternatively, cutting forces may be increased by a factor of several times which, when combined with the relatively low elastic modulus of titanium, can cause serious deflection of the workpiece.

With these considerations in mind it is clear that the machining of titanium must be carried out with high quality, temperature-resistant cutting tools operating at comparatively low speeds. Moreover, because of the tendency of titanium to gall or weld to other metals, sliding contact should be avoided, which means that deep cuts with sharp tools are required. Finally, it is necessary to ensure that both tool and workpiece are rigidly supported.

The above features apply to all titanium alloys although the actual machining conditions vary with different categories. Table 6.5 compares the actual ratios of machining times for CP titanium and typical α, $\alpha + \beta$ and β-alloys with those for the commonly used alloy steel AISI 4340.

The difficulties associated with mechanical machining of titanium and its alloys can be overcome if unwanted metal is removed by chemical dissolution. Thus chemical and, more particularly, electrochemical milling

Table 6.5 Ratios of machining times for various titanium alloys compared with the alloy steel 4340 having a hardness of 300 BHN (from Zlatin, N. and Field, M., In: *Titanium Science and Technology*, Vol. 1. ed. R. I. Jaffee and H. M. Burte, Plenum Press, New York, 1973; p. 409)

Titanium alloy	Hardness (BHN)	Turning (WC tool)	Face milling (WC tool)	Drilling (High speed steel tool)
CP titanium	175	0.7:1	1.4:1	0.7:1
Near-α: Ti-8Al-1Mo-1V	300	1.4:1	2.5:1	1:1
α/β: Ti-6Al-4V	350	2.5:1	3.3:1	1.7:1
β: Ti-13V-11Cr-3Al	400	5:1	10:1	10:1

techniques are sometimes used to produce shapes, the latter employing profiled cathodes.

6.5.5 Joining

Fusion welding The capacity of titanium alloys to be fusion welded is related to microstructure and composition. General weldability is restricted to α, near-α and to α/β alloys containing less than 20% of the β-phase. Fusion, resistance and flash-butt welding have all been used for titanium. Oxyacetylene and atomic hydrogen fusion welding are unsuitable because of gaseous contamination, but non-consumable (TIG) and consumable (MIG) electrode techniques, as well as electron beam and plasma arc processes are applicable. Open air techniques can be used but extra attention must be given to shielding the area being welded with an inert gas such as argon. In this regard it is necessary to take special protective measures, e.g. supplying argon to the underside of the weld, which are not required with aluminium or magnesium alloys. In the case of complex assemblies it is often preferable to weld in a glove-box type of

Fig. 6.19 Polished and etched section of an electron beam butt-weld in 50 mm thick Ti-6Al-4V plate (courtesy Rolls Royce Ltd)

chamber that can be filled with the inert gas. It is important to note that titanium cannot be fusion welded to the other conventional structural materials such as steel, copper, and nickel alloys. Spontaneous cracking or extreme brittleness are inevitable due to the formation of brittle inter-metallic compounds.

Electron beam welding, which is carried out in a vacuum chamber, is especially suitable for titanium alloys although the capital cost of the equipment is high. Specific powers are greater than those usable by other processes so that deep penetration combined with a narrow heat-affected zone is possible, e.g. Fig. 6.19 shows an electron beam butt-weld in 50 mm thick, rolled titanium alloy plate.

Diffusion bonding   Titanium-based materials are also amenable to joining by diffusion-bonding techniques since surface oxide and minor contaminants are readily dissolved at the temperatures used which are usually in the range 850 to 950°C, i.e. below the transus temperature in α/β alloys. It is again necessary to exclude air during bonding and the components being joined are held under low pressures of about 1 MPa for periods of 30 to 60 minutes. Diffusion readily occurs across the interface and it is usual for localized recrystallization to take place there. This is evident in Fig. 6.20 which is a micrograph of a diffusion-bonded joint between two different titanium alloys. Joint strengths are commonly better than 90 % those of the alloys being bonded, providing that the process has no detrimental effect on the microstructure.

Diffusion bonding provides significant cost reductions by eliminating machining operations in the manufacture of shapes having complex

Fig. 6.20   Micrograph of a diffusion-bonded joint between two different titanium alloys. The grain structure shows that localized recrystallization has occurred. The conditions were: temperature 980°C, pressure 1.5 MPa, time 30 min (from Blanchet, B., *Revue de Métallurgie*, **71**, 99, 1974). ×100

geometry. The technique also enables assemblies to be produced in alloys that are not normally regarded as being weldable. As mentioned in Section 6.5.1, diffusion bonding can be combined with superplastic forming in certain fine grained alloys.

Brazing Titanium and its alloys can be brazed at around 1000° C in a protective atmosphere using materials such as silver, copper, and the composition Ti-15Cu-15Ni. An important use of brazing has been in the production of honeycomb structures (see Section 3.7.5) which offer a unique combination of stiffness and corrosion resistance at elevated temperatures.

6.5.6 Powder metallurgy products

Considerable economies in both the use of material and fabricating costs are possible if components can be produced from powders. These techniques are being applied to titanium alloys now that powders of sufficient purity, i.e. particularly low content of interstitial elements, are available. Products can be produced close to final size (near-net shape) with a consequent saving in machining costs.

The two common methods for powder consolidation are hot pressing and sintering and hot isostatic pressing. Sintering is usually carried out in the β-phase field and alloys prepared from elemental powders have been found to sinter more effectively than pre-alloyed powders, probably because of the enhanced diffusion rates at the β-stabilized particle surfaces. Subsequent forging or extrusion to produce components is then carried out in the $\alpha + \beta$ phase field and results in a fine, duplex microstructure. Hot isostatic pressing offers the advantages that pressing, sintering and sometimes forging can be combined in a single operation at temperatures below the $\beta/(\alpha/\beta)$ transus.

Apart from economies in material and fabrication costs, powder metallurgy techniques can offer other advantages. Uniformity of composition is greater because chemical heterogeneity in cast billets is avoided. In addition, components produced from powders show no crystallographic texture or anisotropy of grain shape, and so are much more uniform with respect to mechanical properties. Significantly higher values of proof stress and tensile strength at room and elevated temperatures are also possible, although ductility and toughness are less than that achieved for conventional wrought alloys.

6.6 Titanium alloy castings

The high affinity of molten titanium for oxygen, nitrogen and hydrogen requires that melting and forming be carried out under vacuum. Consumable electrode-vacuum arc or electron beam melting furnaces are commonly used. The range of moulding materials is limited because of the reactivity of titanium and graphite is commonly used for this purpose.

Some precision castings are used in the aerospace industries but the largest castings are prepared for chemical equipment. In the United States spherical valves in sizes up to 2.5 m in diameter and weighing 1 tonne have been produced from titanium alloys. Impellors for pumps is another common application. The two most widely used alloys are Ti-6Al-4V and Ti-5Al-2.5Sn, the latter where welding is required.

6.7 Engineering performance

6.7.1 Tensile and creep properties

The uniquely high strength:weight ratios of titanium alloys over a wide temperature range were shown in Fig. 1.3. Reference has also been made to the fact that the development of titanium alloys has been dominated by the desire to improve creep behaviour at progressively higher temperatures. This theme has been central to the preceding discussion of titanium alloys, with tensile properties and comparative creep behaviour being summarized in Table 6.1 and Fig. 6.2 respectively. In Section 6.2.2, reference was also made to the high specific strength and toughness of α-titanium alloys at very low temperatures which has led to their use for cryogenic storage vessels in space vehicles. Consequently it is not proposed to consider these properties any further.

Recently the near-α alloy IMI 829 was announced, which is available for service for long times at 550 to 600° C. These temperatures are close to the limit at which conventional titanium alloys can be used in air because oxidation normally becomes significant above 600° C. This applies with respect both to surface scaling and internal embrittlement, since oxygen is relatively soluble in titanium above 600° C. There is, however, one class of titanium alloys that shows some promise of being used for products such as gas turbine blades operating at temperatures as high as 900° C. These are the titanium aluminide intermetallic compounds TiAl (Ti-40Al) and Ti_3Al (Ti-16A1) which have a much greater resistance to oxidation. They also have other potential advantages over conventional titanium alloys in terms of relative density and elastic modulus, e.g. 3.6 and 175 GPa respectively, as well as superior creep strength. Their major disadvantages are lack of ductility and poor formability. Both are areas of current research.

The titanium aluminides are produced by powder metallurgy techniques involving pre-alloyed powders which are compacted, extruded to produce rod, and then isothermally hot-forged into shapes. Potential weight savings in aircraft gas turbine engines of 20 % have been predicted if these materials can be successfully exploited, since they have better strength:weight ratios than the nickel or cobalt alloys that are used for components operating above 600° C.

6.7.2. Fatigue behaviour

Titanium and titanium alloys which have isotropic microstructures may display fatigue properties comparable with those obtained with ferrous materials. Curves relating cyclic stress with number of cycles (S/N curves) for smooth specimens tend to show true endurance limits and the ratio of this value to tensile strength is commonly around 0.5. Reductions in this ratio for notched specimens are also characteristic of ferrous materials.

The fatigue performances of titanium and its alloys are, however, very much influenced by microstructure and texture. Effects of microstructure on fatigue behaviour have already been shown in Fig. 6.11 and Table 6.3 whereas the influence of grain direction (and texture) was demonstrated in Fig. 6.18 and Table 6.4. Microstructural effects become more apparent at lower stress levels (high-cycle fatigue) as the plastic zone size ahead of an advancing crack is small and features such as packets of Widmanstätten plates of the α-phase have a greater influence on mode of crack propagation (Section 6.3.1, Fig. 6.11). Fatigue strength is highest in the short transverse direction of wrought products and arises primarily as a result of textures developed within the grains (Section 6.5.3).

A phenomenon which may be unique to certain titanium alloys is the effect of dwell periods at maximum load on rates of growth of fatigue cracks. This effect is shown schematically in Fig. 6.21, and increases in the rate of crack growth of as much as 50 times may occur compared with results obtained in tests on the same material subjected only to sinusoidal stress cycles. Dwell effects are maximized in alloys containing substantial amounts of the α-phase which have a preferred texture such that stressing is normal to the basal planes, whereas they appear to be insignificant if stressing occurs parallel to the basal planes of the α-phase, or if the microstructure is homogeneous and fine grained. Particular attention has been paid to α/β alloys, e.g. Ti-6Al-4V, in which dwell effects have also been found to decrease with increasing amounts of the β-phase in the microstructure. In all cases, dwell effects disappear when stressing occurs at temperatures above 75° C and they are generally considered to arise from the preferential diffusion of hydrogen, during the dwell period, to regions of localized hydrostatic tension ahead of an advancing crack. Such an accumulation of hydrogen would tend to embrittle this region, and it has even been suggested that brittle plates of TiH_2 may be formed.

6.7.3 Corrosion

Titanium is a highly reactive metal but, because of the presence of a very stable, self-healing oxide film, it exhibits excellent corrosion resistance in a wide variety of environments. This is a feature shared with aluminium, although the corrosion resistance of titanium is normally much greater. Titanium is also more resistant to attack than stainless steels and copper alloys in a wide range of conditions, which has led to an expansion of its use by the chemical industry.

The surface oxide film of titanium resists attack by oxidizing solutions, particularly those containing chloride ions which are normally difficult to handle. Titanium shows outstanding resistance to atmospheric corrosion

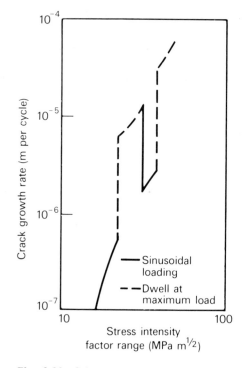

Fig. 6.21   Schematic representation of the effect of dwell periods at maximum load on the rate of crack growth during fatigue tests on certain titanium alloys

in both industrial and marine environments, and its resistance to sea water is virtually unsurpassed by other structural metals. It is also resistant to many acids and salt solutions. However, titanium is attacked in reducing environments in which the oxide film is unstable and cannot be repaired.

One of the most successful applications for titanium alloys is in handling wet chlorine gas, bleaching solutions containing chlorides, hypochlorates and chlorine dioxide. However, titanium suffers catastrophic attack in dry chlorine gas, although as little as 50 parts per million of water will prevent this attack.

For most chemical applications, strength is a secondary requirement and CP titanium is used. In order to cope with non-oxidizing acids such as sulphuric, hydrochloric and phosphoric, or with some reducing conditions, several methods have been developed to improve the corrosion resistance of titanium. An alloy of titanium with 0.2% palladium (IMI 260) was developed in which the role of the noble metal was to induce anodic passivation such that corrosion rates in some solutions have been reduced by a factor of as much as 1000 times. A similar effect may be achieved by external anodic protection in which the potential of titanium is made more positive by connecting to either an electrical power source or to a more noble metal. In this regard titanium has the particular advantage over competing materials such as stainless steel in that it remains passive over a much wider range of potential.

A more recent use of a noble metal as a means of corrosion protection has been the application of the new but rather specialized technique of ion plating to provide a very thin ($\sim 1 \, \mu$m) coating of platinum. Ion plating is a method of coating in which some of the deposited particles are ionized, e.g. by passing through a plasma which assists in cleaning the surface of the substrate and improving the adhesion of the coating. In addition to raising the general corrosion resistance of titanium, ion plating may also increase wear resistance and fatigue strength.

The major corrosion problem with titanium alloys is crevice corrosion which may take place at joints, seams, welds, etc., where circulation of the corroding medium is restricted. Progressive acidification occurs in the crevice because of hydrolysis of corrosion products and the protective oxide film is destroyed. The problem becomes worse at elevated temperatures. Again the addition of the noble metal palladium to titanium has been found to be beneficial, as has ion plating of surfaces with platinum.

6.7.4 Stress-corrosion cracking

The apparent stability and integrity of the oxide film in environments that cause stress-corrosion cracking in more common structural alloys suggested that titanium alloys may be very resistant to this phenomenon. For example, early investigators were unable to crack specimens made from those materials that were stressed and exposed to boiling solutions of 42 % $MgCl_2$ or 10 % NaOH, both of which induce cracking in many stainless steels. However, a susceptibility to this phenomenon was recognized in 1953 with the cracking of CP titanium in red fuming nitric acid, and in 1955 when the unexpected failure of a titanium alloy undergoing a hot tension test was supposedly traced to chloride salts deposited from finger marks. Since then it has been demonstrated in the laboratory that most titanium alloys will undergo stress-corrosion cracking in one or more environments (Table 6.6) and special attention has been given to the comparative performance of alloys in aqueous halides, organic fluids, e.g. methanol, and hot salts. However, it must be emphasized that actual failures in service have been rare.

It can now be appreciated that the reason why titanium alloys were considered to be immune from stress-corrosion cracking in mild environments, and at low temperatures, was the difficulty in initiating cracks. Many titanium alloys are highly resistant to pitting corrosion whereas, in many other materials, the pits provide the stress concentration necessary to initiate stress-corrosion cracking. Although this characteristic of titanium alloys is very desirable, it is now standard practice to use notched or pre-cracked specimens and to apply fracture

Table 6.6 Some environments in which titanium alloys may be susceptible to stress-corrosion cracking (from Boyd, W. K., In: *Proceedings of Conference on Fundamental Aspects of Stress-Corrosion Cracking*, ed. R. W. Staehle *et al.*, Nat. Assoc. Corrosion Eng., 1969; p. 593)

Medium	Temperature (°C)	Examples of susceptible alloys
Cadmium	> 320	Ti-4Al-4Mn
Mercury	Ambient	CP Ti(99%), Ti-6Al-4V
	370	Ti-13V-11Cr-3Al
Silver	470	Ti-5Al-2.5Sn, Ti-7Al-4Mo
Chlorine	290	Ti-8Al-1Mo-1V
Hydrochloric acid		
(10%)	35	Ti-5Al-2.5Sn
	345	Ti-8Al-1Mo-1V
Nitric acid		
(fuming only)	Ambient	CP Ti, Ti-8Mn, Ti-6Al-4V, Ti-5Al-2.5Sn
Chloride salts	290–425	All commercial alloys
Methanol	Ambient	CP Ti(99%), Ti-5Al-2.5Sn, Ti-8Al-1Mo-1V, Ti-6Al-4V, Ti-4Al-3Mo-IV
Trichloroethylene	370	Ti-5Al-2.5Sn, Ti-8Al-1Mo-1V
Sea water	Ambient	CP Ti(99%), Ti-8Mn, Ti-5Al-2.5Sn, Ti-8Al-1Mo-1V, Ti-6Al-4V, Ti-11Sn-2.25Al-5Zr-1Mo-0.25Si, Ti-13V-11Cr-3Al

mechanics techniques when evaluating and comparing the stress-corrosion resistance of different titanium alloys. This follows because of the use of titanium alloys for a number of critical aerospace applications.

It will be apparent from Table 6.6 that stress-corrosion cracking in most alloys is specific to a particular environment. However, it is possible to identify several trends which are related to composition and micro-structure. The compositional features are as follows:

(i) α-titanium alloys generally show the greatest susceptibility to stress-corrosion cracking and even CP titanium will crack in some environments if the level of oxygen in the metal is high. Aluminium contents in excess of 5 to 6% are considered detrimental and this is attributed primarily to the formation of the ordered phase α_2, the presence of which changes the dislocation substructure that forms during deformation. The onset of limited co-planar slip is observed which results in coarse slip steps and localized rupture of the oxide film at the surface, thereby exposing unprotected metal to the corrosive medium. It is proposed that such a mechanism may lead to rapid pitting and the subsequent formation of stress-corrosion cracks, as has been observed in tests on some austenitic stainless steels.

(ii) The addition of elements such as molybdenum and vanadium, which favour formation of the β-isomorphous type of phase diagram, enhance the resistance to stress-corrosion cracking. These elements increase the amount of β-phase and, as this phase appears to be immune from stress-corrosion cracking in a number of α/β alloys, it has been proposed that it serves to arrest cracks that may be propagating in the more susceptible α-phase. It should be noted, however, that elements such as manganese which stabilize β-eutectoid systems increase susceptibility to stress-corrosion cracking.

(iii) Hydrogen is readily absorbed by titanium and is detrimental because a number of alloys are susceptible to hydrogen embrittlement, particularly in aqueous environments. Although the precise mechanism for embrittlement is uncertain, it seems likely to arise from either the formation of brittle plates of titanium hydride, or from the directed diffusion of hydrogen to highly stressed regions such as crack tips, thereby assisting crack propagation. In this regard, it may also be noted that hydride formation occurs most readily in alloys containing aluminium.

Although it is more difficult to propose general rules regarding the role of microstructure, the following trends have been observed:

(i) Quenching from the β-phase field, which often gives acicular micro-structures, generally confers a greater resistance to stress-corrosion cracking than slow cooling from the $\alpha + \beta$ field. This result is similar to that described for fatigue cracking in Section 6.3.1 and is again associated with crack branching, as shown in Fig. 6.11b. However, it should be noted that the reverse seems to hold for alloys exposed to halides at high temperatures, i.e. so-called 'hot salt stress-corrosion cracking'.

(ii) A combination of plastic deformation and heat treatment is usually beneficial in reducing susceptibility because it refines the microstructure and reduces grain size.

(iii) Ageing leading to precipitation of a second phase, e.g. ω or α, within β-grains can lower resistance to stress-corrosion cracking.

(iv) As with aluminium alloys, wrought titanium products show greatest susceptibility to stress-corrosion cracking in the short transverse direction.

6.7.5 Corrosion-fatigue

The combination of the generally high fatigue strength of titanium alloys and their good resistance to corrosion suggests that they should perform well under conditions of corrosion-fatigue and this has been found to be so. Special attention has been given to comparisons between the α/β alloy Ti-6Al-4V and 12% chromium steel, because of the possible replacement of this material by titanium alloys for the low pressure sections of large steam turbines. The strength level at which titanium alloys could be put into service is some 25% higher than that for a 12% chromium steel. Taking into account this factor and the difference in relative density, the endurance limit for Ti-6Al-4V tested in air is more than double that for the steel (Fig. 6.22). Moreover, whereas the titanium alloy is unaffected by testing in steam or in a solution of 3.5% NaCl, the endurance limit for the steel fatigue tested in only 1% NaCl solution falls to approximately one-eighth that of the titanium alloy.

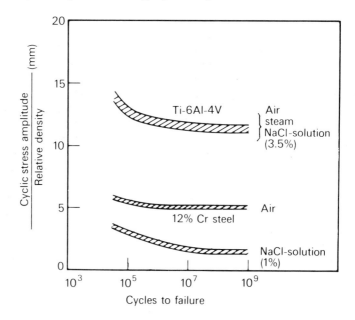

Fig. 6.22   Fatigue properties (compensated for differences in relative density) of Ti-6Al-4V and a 12% chromium steel tested in different environments (from Jaffee, R. I., *Met. Trans.*, **10A**, 139, 1979)

6.8 Applications of titanium alloys

6.8.1 Aerospace

Although the development of titanium alloys only commenced in the late 1940s, their uniquely high strength:weight ratios led to their introduction in aircraft gas turbines as early as 1952 when they were used for compressor blades and discs in the famous Pratt and Whitney J57 engine. Immediate weight savings of approximately 200 kg were achieved. At the same time, in Britain, an alloy Ti-2Al-2Mn was used for the equally famous Rolls Royce Avon engine that powered such aircraft as the Comet and Canberra. Since then, aircraft gas turbines have continued to provide the major application for titanium alloys which now make up 25 to 30 % of the weight of most modern engines. In particular, titanium alloys have played a major role in by-pass (fan-jet) engines for which a large front fan is required (Fig. 6.9). In addition to the fan, titanium alloys are used for most of the blades and discs in the low and intermediate sections of the compressors of modern engines as shown, for example, in Fig. 6.23. In general the α/β alloy Ti-6Al-4V is favoured for the fan and cooler parts of the compressor, whereas near-α alloys are specified when greater creep resistance is required. Other engine applications include the use of titanium sheet for casings and ducting.

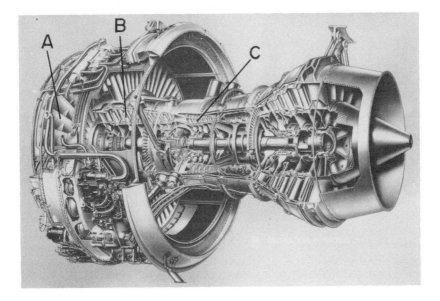

Fig. 6.23 Sectional view of the Rolls Royce RB 211 gas turbine engine. A = fan blades, B = low pressure compressor, C = intermediate pressure compressor (courtesy Rolls Royce Ltd)

The use of titanium alloys for structural members in aircraft developed more slowly because of their high cost compared with aluminium alloys. They were first used mainly in the form of sheet for nacelles, exhaust shrouds, and fire walls where heating is significant. They were, however, used to a much greater extent for the structures of several space vehicles and for fuel storage tanks in rockets. An increasing use is now being made of forgings for quite critical applications, including undercarriage components, flap and slat tracks in wings and for engine mountings. For example, the Boeing 707 which first came into service in 1958 contained only 80 kg of titanium alloys, the Boeing 727 (1963) 290 kg, the Boeing 747 (1969) 3850 kg, and the McDonnell-Douglas DC 10 (1971) 5500 kg, which represents more than 10 % of the structural weight. Considerable use is also being made of high-strength titanium alloy fasteners and the large American military transport aircraft, the Lockheed C 5A, uses 1.5 million units out of a total of 2.2 million. This has resulted in a direct reduction in weight of 1 tonne, with an additional 3.5 tonnes being saved through consequent structural modifications that are possible by using titanium alloy fasteners.

Some of the high costs associated with titanium alloys have arisen because of the use of fabrication techniques developed for aluminium. More recently, significant reductions in both costs and weight have been achieved by the adoption of novel forming and joining techniques for

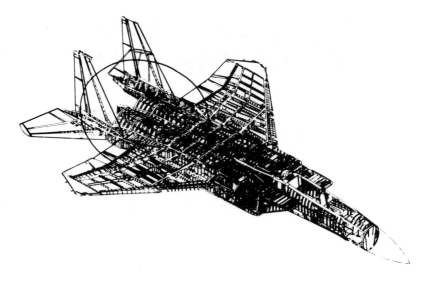

Fig. 6.24   A structural diagram of the McDonnell-Douglas F15 aircraft in which the encircled area is made almost entirely from titanium alloys (from Tupper, N. G. *et al.*, *J. Metals*, **30**, No. 97, 1978)

titanium, some of which were briefly described in Section 6.5. As a result, the amount of titanium alloys incorporated in the structures of some of the latest aircraft has been further increased. For example the McDonnell-Douglas F15 fighter aircraft contains some 7000 kg of titanium alloys which represents 34 % of the structural weight as compared with 48 % for aluminium alloys. A sketch of the F15 structure is shown in Fig. 6.24 and the encircled area is made almost entirely from titanium alloys.

6.8.2 Chemical and general engineering

Use of titanium in these areas amounts to approximately 15 % of production in the United States and 40 % in Europe, and these proportions seem likely to increase. Contrary to aerospace, most other engineering applications have involved the use of CP titanium so that comparatively little alloy development has been needed. A notable exception is the alloy Ti-0.2Pd (IMI 260).

As discussed in Section 6.7.3, it is the outstanding corrosion resistance of titanium in many environments that is the prime reason for its use in chemical and general engineering. Resistance to chlorides is perhaps its most valuable attribute and titanium is uniquely suited for equipment handling solutions, e.g. bleaching agents, hot brine, ferric chloride and chlorinated hydrocarbons. Another major application is for tubing in steam condensors and heat exchangers, with the offshore oil rigs providing

Fig. 6.25 Tubing and baffles for gas/oil product coolers fabricated from titanium tube and plate (from *Metallurgist and Materials Technologist*, **9**, 543, 1977)

classic conditions for using titanium (Fig. 6.25) since sea water is used as a coolant and the oil and gas contain sulphide contaminants. Titanium has a notably higher corrosion and erosion resistance than the copper alloys which it replaces so that thinner, e.g. 0.5 mm, walled tubing can be used. This improves heat transfer despite the relatively poor thermal conductivity of titanium.

An area of likely future expansion in the use of titanium is for blading in the low pressure section of large steam turbines. For many years these blades have been made from a 12 % chromium steel and their size has been limited by the centrifugal loads imposed on the supporting rotors. As shown in Fig. 6.22, Ti-6Al-4V has much better corrosion-fatigue properties than the steel and, for comparable stresses on the rotors, titanium alloy blades can be some 40 % longer. Some blades are now operating which exceed 1 m in length. Erosion by wet steam is a particular problem in this section of turbines and, in this regard, titanium alloys are also superior to the 12 % chromium steels although not quite as good as the stellite shield that is commonly bonded to the leading edge of the steel blades. The major disadvantage of titanium alloys is their reduced stiffness arising from a lower elastic modulus, and more rigid positioning is necessary to compensate for this. The current cost differential for titanium alloy blades is predicted to be 40 to 50 % higher and it is unlikely that stellite erosion shields will be required.

6.8.3 Biomaterials

Titanium and titanium alloys show an outstanding resistance to corrosion by body fluids which is superior to that of stainless steels. This factor,

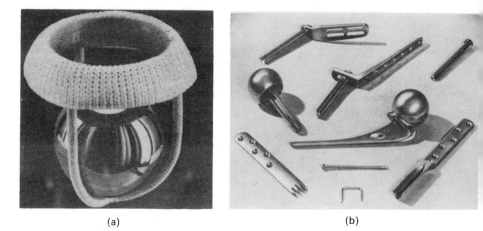

(a)   (b)

Fig. 6.26 Prosthetic devices made from titanium and titanium alloys: a, Starr-Edwards heart valve; b, artificial joints, plates and pins

together with good stress-corrosion resistance, high mechanical properties and acceptable tissue tolerance has led to their use for prosthetic devices. A notable example has been the Starr-Edwards heart valve in which a dacron covered titanium cage contains a hollow, electron beam welded titanium ball (Fig. 6.26a). It should be noted that this ball has been designed to have a relative density similar to blood and it replaces a heavier silicone ball which was less satisfactory due to inertial effects. Fig. 6.26b shows a number of other prostheses including plates, pins and an artificial hip joint.

Further reading

McQuillan, A. D. and McQuillan, M.K., *Titanium*, Butterworths, London, 1956

Zwicker, U., *Titanium and Titanium Alloys*, Springer-Verlag, Berlin, 1974 (in German)

Jaffee, R. I. and Promisel, N. E. (eds), *The Science, Technology and Applications of Titanium Alloys*, Proceedings of the First World Conference on Titanium, Pergamon, London and New York, 1970[†]

Jaffee, R. I. and Burte, H. M. (eds), *Titanium Science and Technology*, Proceedings of the Second World Conference on Titanium, Plenum Press, New York and London, 1973[†]

Hammond, C. and Nutting, J., The physical metallurgy of superalloys and titanium alloys, *Metal Science*, **11**, 474, 1977

Williams, J. C., In: *Precipitation Processes in Solids*, ed. K. C. Russell and

[†] Proceedings of the Third World Conference on Titanium (Russia, 1976) were published in Russian in 1978, and those of the Fourth Conference (Japan, 1980) will be published in English in 1981.

H. L. Aaronson, Met. Soc. AIME, New York, 1978; Chapter 7

McQuillan, M. K., Phase transformations in titanium alloys, *Met. Reviews*, **8**, 41, 1963

Miska, K. H., Titanium and its alloys, *Mater. Eng.*, Manual 48, **80**, No. 1, 61, 1974

Goosey, R. E., Heat treatment of engineering components, *The Iron and Steel Institute Publication No. 124*, 1969; p. 75

Hickman, B. S., Formation of omega-phase in titanium and zirconium alloys: a review, *J. Materials Science*, **4**, 554, 1969

Hall, I. W. and Hammond, C., Fracture toughness and crack propagation in titanium alloys, *Mater. Sci. Eng.*, **32**, 241, 1978

Forsyth, P. J. E. and Stubbington, C. A., Directionality in structure-property relationships: aluminium and titanium alloys, *Met. Technology*, **2**, 158, 1975.

Larson, F. and Zarkades, A., Properties of textural titanium alloys, *Metals and Ceramics Information Center Report MIC-74-20*, Watertown, Massachusetts, 1974

Stubbington, C. A. and Pearson, S., Effect of dwell on the growth of fatigue cracks in Ti-6Al-4V bar, *Eng. Fracture Mech.*, **10**, 723, 1978

Boyd, W. K., In: *Handbook on Stress Corrosion Cracking and Corrosion Fatigue*, ed. R. W. Staehle and M. O. Speidel, Corrosion Center, Ohio State University, Ohio, 1980 (in press)

de Gelas, B. *et al.*, Mechanical properties, resistance to corrosion, and the uses of titanium in the chemical industry, *Rev. Met.*, **71**, 75, 1971

Kuhlman, G. W. and Billman, F. R., Selecting processing options for high-fracture toughness titanium airframe forgings, *Met. Prog.*, **111**, No. 3, 39, 1977

Morton, P. H., Titanium alloys for engineering structures, *Rosenhain Centenary Conference on the Contribution of Physical Metallurgy to Engineering Practice*, The Royal Society, London, 1976

Farthing, T. W., Introducing a new material – the story of titanium, *Proc. Inst. Mech. Eng.*, **191**, 59, 1977

Froes, F. H. *et al.*, Developments in titanium powder metallurgy, *J. Metals*, **32**, No. 2, 47, 1980

Appendix

Table A.1 Unit conversion factors

Property	To convert B to A multiply by	SI units (A)	Non-SI units (B)	To convert A to B multiply by
Mass	0.4536	Kilogram (kg)	Pound (lb)	2.204
	0.4536×10^{-3}	Tonne	lb	2204
	1.0163	Tonne	UK ton	0.9839
Stress	6.894×10^{-3}	Megapascal (MPa) (Meganewtons per square metre)	Pounds force per square inch (psi)	145.04
	15.444	MPa	UK tons force per square inch (tsi)	6.475×10^{-2}
	9.8065	MPa	Kilograms per square millimetre (kg mm^{-2})	0.10197
Fracture toughness	1.0989	Megapascal $(metre)^{1/2}$ (MPa $m^{1/2}$)	Kilopounds force per square inch $(inch)^{1/2}$ (ksi $in^{1/2}$)	0.91004
Thermal conductivity	4.1868×10^{2}	Watts per metre per degree Kelvin	Calories per centimetre per second per degree Centigrade	2.3885×10^{-3}
Specific heat capacity	4.1868×10^{3}	Joules per kilogram per degree Kelvin (J $kg^{-1}K^{-1}$)	Calories per gram per degree Centigrade	2.3885×10^{-4}

Index